Erwin Dee Kord (Ed.)

Sri Lanka Artillery

Erwin Dee Kord (Ed.)

Sri Lanka Artillery

Regiments and Corps of the Sri Lanka Army

Solv

Imprint

All parts of this book are extracted from Wikipedia, the free encyclopedia (www.wikipedia.org).

You can get detailed informations about the authors of this collection of articles at the end of this book. The editors (Ed.) of this book are no authors. They have not modified or extended the original texts.

Pictures published in this book can be under different licences than the GNU Free Documentation License. You can get detailed informations about the authors and licences of pictures at the end of this book.

The content of this book was generated collaboratively by volunteers. Please be advised that nothing found here has necessarily been reviewed by people with the expertise required to provide you with complete, accurate or reliable information. Some information in this book maybe misleading or wrong. The Publisher does not guarantee the validity of the information found here. If you need specific advice (f.e. in fields of medical, legal, financial, or risk management questions) please contact a professional who is licensed or knowledgeable in that area.

Cover image: www.ingimage.com
Concerning the licence of the cover image please contact ingimage.

Publisher:
Solv is a trademark of
International Book Market Service Ltd., 17 Rue Meldrum, Beau Bassin, 1713-01 Mauritius
Email: info@bookmarketservice.com
Website: www.bookmarketservice.com

Published in 2012

Printed in: U.S.A., U.K., Germany. This book was not produced in Mauritius.

ISBN: 978-613-9-18035-6

Contents

Articles

References

Sri_Lanka_Artillery

Sri Lanka Artillery	
Active	12 April 1888 - Present
Country	Sri Lanka
Branch	Sri Lanka Army
Type	Artillery
Role	Artillery, Target Acquisition, Ceremonial
Size	12 regular regiments, 2 volunteer regiments
Part of	Artillery Brigade
Regimental Headquarters	Panagoda Cantonment, Panagoda.
Nickname	SLA
March	*The British Grenadiers*
Anniversaries	12 April
Engagements	World War I World War II 1971 Insurrection Insurrection 1987-89 Sri Lankan Civil War
Commanders	
Centre Commandant	Col A S Hewawitharana IG
Colonel-Commandant	Maj Gen J.C.Rambukpotha RSP USP ndc IG
Notable commanders	Colonel F. C. de Saram OBE

The **Sri Lanka Artillery (SLA)** is a corps of the Sri Lanka Army. It is made up of 7 regular regiments and 2 volunteer (reserve) regiments. The SLA is headquartered at Panagoda Cantonment, Panagoda.

History

The roots of the Sri Lanka Artillery goes back to 1888, when on 12 April 1888 the **Ceylon Artillery Volunteers** was formed as a gun battery under the command of Capt. C.E.H Seimans of the Royal Artillery. During World War I the Ceylon Artillery Volunteers were mobilised, along with the **Town Guard Artillery** of Colombo for the defence of Colombo. In 1918 the **Ceylon Garrison Artillery** was formed by amalgamating the Ceylon Artillery Volunteers and Town Guard Artillery, coming under the Ceylon Defence Force. During the colonial period the main responsibility of the Ceylon Garrison Artillery was the coastal defence of Ceylon, by manning Coastal artillery batteries.

During World War II the CGA was deployed outside Ceylon to support the Allied and Commonwealth forces in the Indian Ocean, as well as defending Ceylon. Equipped with 6 inch Guns the CGA was deployed in defence of the Seychelles and the Cocos Islands. Within Ceylon expanded in size, with the 1st Coast Regiment raising in strength of one field and four coast batteries, the formation the 2nd Heavy Anti-Aircraft Regiment, headquartered in Trincomalee and the 3rd Searchlight/Light Anti-Aircraft Regiment based in Colombo.[1]

In 1949 the CGA became the **Ceylon Artillery** with the formation of the Ceylon Army under the Army Act of 1949. That year the 1st Heavy Anti-Aircraft Regiment (HAA) of the Ceylon Artillery was formed under the command of Lt. Col. (later Colonel) F. C. de Saram OBE. In 1953 it was renamed the 1st Light Anti-Aircraft Regiment (LAA). In the same year the 3rd Field Artillery Regiment was formed. The 2nd Volunteer Coastal Artillery / Anti-Aircraft Regiment was formed in 1949, under the command of Lt. Col. J.A.T. Perera ED. It was made up of officers and men of the Ceylon Garrison Artillery. In 1962, when many officers of the CA, including Colonel De Saram were frond associated with the attempted military coup the government began rapid reduction of the CA. As a result in 1962 2nd Volunteer Regiment was disbanded and coastal artillery batteries decommissioned. The following year in 1963 the 1st Heavy Anti-Aircraft Regiment was amalgamated with 3rd Field Artillery Regiment and the 4th Regiment, Ceylon Artillery was formed, thus reducing the Ceylon Artillery to a single regiment. Years after amalgamation of the LAA, the air defence role of the Sri Lankan military was taken over by the Sri Lanka Air Force Regiment which it carries out to this day. The CA was renamed in 1972 as the **Sri Lanka Artillery** when Ceylon became a republic. Only in 1980 was a new volunteer regiment, the 5th (V) Artillery Regiment was formed.

At present the Sri Lanka Artillery has 12 regular regiments, 2 volunteer regiments and a regimental band. These units form the Artillery Brigade. Since the start of the Sri Lankan civil war in the 1980s, the SLA has provided fire support for almost all military operations carried out by the Sri Lanka Army against terrorist forces. It also performs a target acquisition role to locate enemy guns. At times the SLA has deployed 4 regular (RFT)and 2 volunteer regiments in an infantry role to meet the shortage of infantry.

Units

Regular Regiments

- 4th Field Regiment SLA
- 6th Field Regiment SLA
- 7th Light Regiment SLA
- 8th Field Regiment SLA
- 9th Field Regiment SLA
- 10th Field Regiment SLA
- 11th Regiment[RFT] SLA
- 14th Rocket Regiment SLA
- 15th Field Regiment SLA
- 16th Regiment[RFT] SLA
- 17th Regiment[RFT] SLA
- 18th Regiment [RFT] SLA
- School Of Artillery [2]

Volunteer Regiments

- 5th(v) Sri Lanka Artillery Regiment
- 12th(v) Sri Lanka Artillery Regiment

Equipment

A RM-70 multiple rocket launcher of the Sri Lanka Artillery.

During the colonial period the Ceylon Garrison Artillery personnel manned British Coastal artillery batteries. After independence and establishment Artillery Regiment, the corps depended on its World War II-era British BL 6 inch coastal guns, 40 mm anti-aircraft guns, 3.7 inch heavy anti-aircraft guns and 4.2-inch heavy mortars. Later British Ordnance QF 25 pounder Mark III field guns and 76 mm mountain guns where introduced.

New sources of weaponry in the 1990s and 2000s included China and the Czech Republic. With the escalation of the Sri Lankan civil war 120 mm, 152 mm, 130 mm howitzers and 120 mm, 82 mm mortars where introduced and in 2000 RM-70 Multiple rocket launchers where added increasing SLA's fire power. Target acquisition systems are also operated by the corps to locate enemy guns. The air defence role has now been completely taken over by the Sri Lanka Air Force Regiment.

Multiple rocket launchers

- RM-70 Multiple rocket launcher
- BM-21 Multiple rocket launcher

Artillery

- 152 mm Type 66 gun-howitzers
- 130 mm Type 59 field guns
- 122 mm Type 83 howitzers
- 122 mm Type 54 howitzers
- 85 mm field guns

Artillery (Ceremonial)

- Ordnance QF 25 pounder field guns - (Ceremonial Gun Troop)
- Ordnance QF 75 mm mountain guns - (Ceremonial Gun Troop)

Heavy and Towed mortars

- Type 86 (W86) 120 mm mortars [3]
- Type 84 (W84) 82 mm mortars [4]

Notable members

- General C. S. Weerasooriya RWP, RSP, VSV, USP, SLA - Former Commander of the Army.
- Lieutenant General Hamilton Wanasinghe VSV, SLA - Former Commander of the Army.
- General L. P. Balagalle VSV, USP, ndc, IG, SLA - Former Commander of the Army.
- Major General Jayantha de S. Jayaratne, VSV, ndc, IG, SLA - former Commander Security Forces Jaffna
- Major General Sarath Munasinghe RWP, RSP, USP - Former Deputy Speaker of Parliament and Commander Security Forces Headquarter Jaffna.
- Major General Larry Wijeratne † - Former Brigade commander, 514 Brigade.
- Major General Ananda Hamangoda † - Former Brigade commander, 512 Brigade.
- Brigadier Bhathiya Jayatilleka, RSP, SLA † - former Brigade commander, 54-1 brigade
- Colonel Fredrick C. de Saram OBE, CA - former Commanding officer, Ceylon Artillery & Leader of the Attempted military coup in 1962
- Lieutenant Colonel Wilmot "Willie" S. Abrahams, MC, MBE, CA - former Commanding Officer, 3rd Field Artillery Regiment, Ceylon Artillery & accused conspirator in the 1962 coup d'état attempt
- Major Victor Gunasekara, CCS - former Controller of Imports Exports and Government Agent of Kegalle
- Second Lieutenant Ajantha Mendis - National Cricketer[5]
- Bombardier Gratien Fernando - Leader of the Cocos Islands Mutiny

Alliances

- United Kingdom - Royal Regiment of Artillery
- India - Regiment of Artillery

Order of precedence

See also

Sri Lanka Army

References

[1] The British Armies in World War Two: An Organisational History, vol 9: The Indian Army, part two: The Indian Army in the East, 1939-43 (http://stonebooks.com/archives/070121.shtml)

[2] http://www.army.lk/arty.php

[3] http://www.sinodefence.com/army/crewserved/type86mortar_120mm.asp

[4] http://www.sinodefence.com/army/crewserved/type84mortar_82mm.asp

[5] Soldier creating history in International cricket (http://www.defence.lk/new.asp?fname=20080613_04)

External links and sources

- Sri Lanka Army (http://www.army.lk)
- Sri Lanka Artillery (http://www.gunner.lk)
- The Cocos Mutiny, 66 years on, as recalled by (Donny) Vincent Ranasinghe, the last surviving member of the Ceylon Garrison Artillery (CGA) who served on the island (http://www.sundaytimes.lk/080420/Plus/plus00001.html)

Sri_Lanka_Army

Sri Lanka Army	
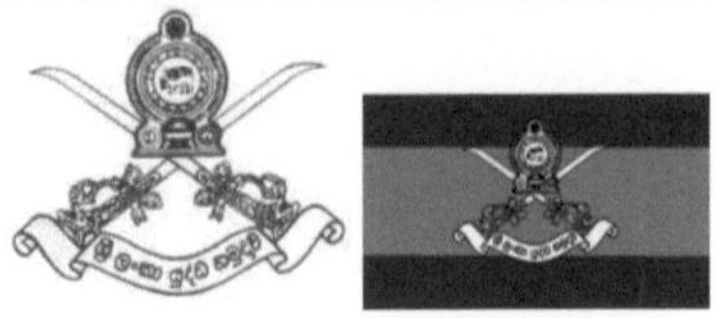 Sri Lanka Army Insignia and Flag	
Active	1 April 1881 - Present [1]
Country	Sri Lanka
Branch	Sri Lanka Armed Forces
Type	Army
Role	Foreign and Domestic Defense
Size	276,700[2]
Part of	Ministry of Defence
Military Headquarters	Army Headquarters, Colombo
Motto	Latin: *Pro Patria* "For Country"
Anniversaries	October 9
Engagements	World War II 1971 JVP Insurrection 1987–89 JVP Insurrection Sri Lankan Civil War
Commanders	
Commander of the Army	Lt.Gen. Jagath Jayasuriya
Notable commanders	Gen. D.S. Attygalle Gen. Sarath Fonseka Lt. Gen. Denzil Kobbekaduwa Maj. Gen. Vijaya Wimalaratne Maj. Gen. Bertram Heyn Maj. Gen. Anton Muttukumaru

Sri Lanka Army

Components

Sri Lanka Army

Sri Lanka Army Volunteer Force

List of regiments and corps

Structure of the Sri Lanka Army

History

History of the Sri Lanka Army

Equipment

Weaponry and equipment

Personnel

Commander of the Army

Army Ranks

Sri Lankan Civil War

(1983–2009)

Background

Sri Lanka · History of Sri Lanka

Origins of the Civil War

Origins of the Civil War · Sri Lankan Tamil nationalism · Sinhalese Buddhist nationalism · Riots · Black July

Main phases

Eelam War I · Indian intervention · Eelam War II · Eelam War III · Eelam War IV

LTTE

LTTE (Divisions) · Black Tigers · Attacks · Expulsion of Muslims · Suicide bombings

Military of Sri Lanka

Military · Civilian attacks · Army · Navy · Air Force · Police · Home Guards · LRRP · STF

Major leaders

M. Rajapaksa · V. Prabhakaran · C. Kumaratunga · A. Balasingham · J. R. Jayewardene · Karuna

Indian involvement

Operation Poomalai · Indo-Sri Lanka Accord · Indian Peace Keeping Force · Operation Pawan · Rajiv Gandhi · RAW

Militant & paramilitary groups

Tamil militant groups (List) · ENDLF · ENLF · EPDP · EPRLF · EROS · PLOTE · TELO · TMVP

Other

Battles · Casualties · War crimes · LLRC · State terror · Human rights · Disappearances · Child soldiers · Assassinations · Protests (Canada)

The **Sri Lanka Army** is the oldest and largest of the Sri Lanka Armed Forces and is responsible for land-based military and humanitarian operations. Established as the **Ceylon Army** in 1949, it was renamed when Sri Lanka became a republic in 1972. As of the year 2010, the Army has believed to have approximately 200,000 regular personnel, approximately between 20,000-40,000 reserve personnel and 18,000 National Guardsmen[3] [4] and comprises 13 operational divisions, one air-mobile brigade, one commando brigade, one special forces brigade, one independent armored brigade, three mechanized infantry brigades and over 40 infantry brigades.[5] From the 1980s to 2009 the army was engaged in the Sri Lankan civil war.

The professional head of the Sri Lanka Army is the Commander of the Army, currently Lieutenant General Jagath Jayasuriya.[6] The Commander-in-Chief of the Sri Lankan Military is the President of the country, who heads the National Security Council through the Ministry of Defense, which is the highest level of military command charged with formulating, executing defence policy and procurements for the armed forces.[7] However operations of the Sri Lanka Army are coordinated by the Joint Operations Command, with other two armed forces.

The Sri Lankan Army has also provided special training to the United States Army on their request as well as many other countries in military education regarding civilian rescue, jungle combat, and guerilla warfare etc.[8]

History

Ancient and pre-colonial times

The first military engagements in Sri Lankan history were marked by the advent of King Vijaya, a Bengal prince who landed along with his followers on the beaches of northwestern Sri Lanka around 543 BC. Prince Vijaya and his followers occupied the lands of the native Veddah people. Repeated incursions by South Indians, particularly the Cholas, into Sri Lankan territory occurred throughout the next few centuries and led to the engagement of the rival forces in battle.[9] In one famous encounter, Sinhalese King Dutugemunu (200 BC) raised an army of eleven thousand inhabitants in his battle against the Chola invader King Elara, whom he eventually defeated. King Dutugemunu's organizational skills, bravery and chivalry are famous and his battles have gone down in history as outstanding offensive operations.[10]

Other Sri Lankan rulers whose military achievements stand out include King Gajabâhu (113 AD), who sailed to India to bring back his captured soldiers, and King Dhatusena (433 AD) who is credited with repulsing numerous Indian invasions and for organizing a naval build-up to deter seaborne attacks. He also had the foresight to cover his defenses with artillery. Vijayabâhu I (1001 AD) was another warrior king who dislodged Indian invaders and united the country. Parakramabahu the Great (1153 AD) was an outstanding monarch of the Polonnaruwa period of Sri Lankan history, and his accomplishments as a military leader and a great administrator are noteworthy. His reign included a military expedition to Burma (Myanmar) in retaliation for indignities inflicted on his envoys and Burmese interference in the elephant trade. This marked the first overseas expedition in Sri Lankan military history. It is also reported that Parakramabahu's fame was such that his assistance was sought by South Indian rulers who were involved in internecine struggles. Another strong ruler in the pre-colonial era was Parâkramabâhu VI, who defeated Indian invaders, united the island and ruled it from capital Sri Jayawardhanapura, Kotte.[10] Although the known epigraphical records do not indicate that the Sri Lankan rulers had a full-time standing army at their disposal, there is evidence supported by legend, designation, name, place and tradition that prove there were 'stand-by' equestrian, elephant, and infantry divisions to ensure royal authority at all times. Militias were raised as the necessity arose, and the soldiers returned to their pursuits, mainly for farming, after their spell of military duty.[10]

Colonial era

Parts of Sri Lanka came under the control of three colonial European powers, namely the Portuguese in the 16th century, the Dutch in the 17th century and the British in the 18th century. Yet, until the entire island was ceded to the British in 1815, regional kingdoms maintained most of their independent defense forces and were able to successfully repulse repeated thrusts by the European armies. However the British, unlike their counterparts, were not primarily restricted to maritime power, and thus had the capability to bring the entire island under their control and to integrate locals into the British defense forces.[10]

Portuguese and Dutch rule (1505-1796 AD)

In the beginning of the 16th century, modern Europe first came in contact with Sri Lanka, which was then referred to as Ceylon. In 1505 a Portuguese fleet, while operating in the Indian seas against Arab traders, was blown off course and landed at Galle, on the southern coast of the island.[11] In 1517 the Portuguese re-appeared, and with the consent of the Sinhalese King established a trading post in Colombo. Having initiated contact with Sri Lanka as traders, the Portuguese soon made themselves political masters of the western seaboard. Numerous forts were soon established, and features of European civilization were introduced.[10]

The Portuguese are credited with the introduction of European-style fortresses to Sri Lanka during this era. Although some locals already possessed military training and fighting experience, there is no evidence that the Portuguese employed local inhabitants into their own forces. Thus the Portuguese were forced to restrict their presence in the island due to their small numbers and their efforts were more focussed toward projecting maritime power.[10]

In 1602 Dutch explorers first landed in Sri Lanka, which was then under Portuguese control. By 1658 they had completely ousted the Portuguese from the coastal regions of the island. Much like the Portuguese, they did not employ locals in their military, and preferred to live in isolation, pursuing their interests in trade and commerce. Like the Portuguese, they defended their forts with their own forces, but unlike the Portuguese, Dutch forces employed Swiss and Malay mercenaries. The Dutch Forts in Jaffna, Galle, Matara, Batticaloa and Trincomalee were sturdily built and are considered a tribute to their military engineering skills. Also, like the Portuguese, the Dutch focussed on maritime power and although they had the capability to develop and use local forces, they chose to isolate themselves from the local population.[10]

British rule (1798-1948 AD)

The British Empire then ousted the Dutch from the coastal areas of the country, and sought to conquer the independent Kandyan Kingdom. In the face of repeated British assaults, the Kandyans were forced into a degree of guerrilla warfare and fared well against their superior British adversaries.[10]

Initially the British stationed their forces, which included naval vessels, artillery troops and infantry, to defend of the island nation from other foreign powers, using the natural harbor of Trincomalee as their headquarters in Sri Lanka. In 1796, the Swiss and Malay mercenaries who were previously in the service of the Dutch were transferred to the British East India Company. While the Swiss De Meuron's Regiment was eventually disbanded in Canada in 1822, the Malays, who initially formed a Malay Corps, were converted into the 1st Ceylon Regiment in 1802 and placed under a British commanding officer. In the same year, the British became the first foreign power to raise a Sinhalese unit, which was named the 2nd Ceylon Regiment, also known as the Sepoy Corps.[10]

In 1803 the 3rd Ceylon Regiment was created with Moluccans and recruits from Penang. All these regiments fought alongside British troops in the Kandyan wars which began in 1803. Throughout the following years, more Sinhalese and Malays were recruited to these regiments, and in 1814 the 4th Regiment was raised, which was composed entirely of African troops. It was later renamed as the Ceylon Rifle Regiment. Eventually, the Kandyan Kingdom was ceded to the British in 1815, and with that they gained control over the whole island. Resistance to British occupation cropped up almost instantly. During the first half-century of occupation, the British faced a number of uprisings, and were forced to maintain a sizable army in order to guarantee their control over the island. After the Matale Rebellion led by Puran Appu in 1848, in which a number of Sinhalese recruits defected to the side of the rebels, the recruitment of Sinhalese to the British forces was temporarily halted.[10]

Ceylon Light Infantry Volunteers

The second phase in the employment of non-British personnel commenced in 1861 after the enactment of an ordinance designed to authorize the creation of a Volunteer Corps in the island. It was designated the Ceylon Light Infantry Volunteers (CLIV). This move compensated for the disbandment of the Ceylon Rifle Regiment in 1874. The Ceylon Light Infantry Volunteers was originally administered as a single unit. However, over the years various sections of the volunteers grew large enough to become independent from their parent unit. The different units that emerged from the Volunteer Force were the

- Ceylon Artillery Volunteers
- Ceylon Mounted Infantry (CMI)
- Ceylon Volunteer Medical Corps
- Cadet Battalion Ceylon Light Infantry
- Ceylon Engineers
- Ceylon Supply & Transport Corps
- Ceylon Planters Rifle Corps (CPRC).[10]

Ceylon Defence Force

In 1910 the name of the military was formally changed to the Ceylon Defence Force (CDF). It continued to grow throughout the early period of the 20th century. The CDF saw active service when a contingent of the Ceylon Mounted Infantry (CMI) in 1900, and a contingent of the Ceylon Planters Rifle Corps (CPRC) in 1902, took part in the Second Boer War in South Africa. Their services were recognized by the presentation in 1902 of a color to the CMI, and a presentation in 1904 of a banner to the CPRC. In 1922, the CDF was honored by the presentation of the King's and Regimental colors to the Ceylon Light Infantry (CLI).[10]

First Prime Minister of Independent Sri Lanka Hon. D.S.Senanayaka visiting the 1st battalion of the CLI at the Echelon Square and watching volunteers being trained to handle light machine guns.

During the First World War, many volunteers from the Defence Force traveled to England and joined the British Army, and many of them were killed in action. One of them mentioned by Sir Arthur Conan Doyle was Private Jacotine of the CLI, who was the last man left alive in his unit at the Battle of Lys, and who continued to fight for 20 minutes before he was killed.[12]

In 1939, the CDF was mobilized and an enormous expansion took place which required the raising of new units such as the Ceylon Signals Corps, the Auxiliary Territorial Service (Ceylon) and also the Colombo Town Guard, which had been previously disbanded, but was later re-formed to meet military requirements. During the Second World War, Britain assumed direct control over the Armed Forces of Ceylon.[13]

Post-independence

Brigadier James Sinclair, Earl of Caithness inspecting a guard of honour wearing khaki drill.

At the end of World War II, CDF which had increased in size during the war began demobilization.In 1948 Sri Lanka gained independence from Britain, becoming a Dominion within the commonwealth and a year earlier Ceylon entered into the bi-lateral Anglo-Ceylonese Defence Agreement of 1947. This followed by the Army Act No. 17 of 1949 which was passed by Parliament on April 11, 1949 and formalized in Gazette Extraordinary No. 10028 of October 10, 1949 marked the creation of the Ceylon Army, consisting of a regular and a volunteer force, the later being the successor of the disbanded CDF.[14] [15] Therefor October 10, 1949 is considered the Ceylon Army was raised, and October 10 is celebrated annually as Army day. The Defence Agreement of 1947 provided the assurance that British would come to the aid of Ceylon in the event it was attacked by a foreign power and provided British military advises to build the country's military. Brigadier James Sinclair, Earl of Caithness was appointed as general officer commanding Ceylon Army, as such becoming the first commander of the Ceylon Army.

The initial requirement was to raise an artillery regiment, an engineer squadron, an infantry battalion, a medical unit, and a service corps company. For much of the 1950s the army was preoccupied with the task of building itself and training existing and new personal. To this aim the British Army Training Team (BATT) advisory group carried out training for ex-members of the CDF within the Ceylon Army, senior officers were sent to the British Army Staff College, Camberley and some attached to units of the British Army of the Rhine to gain field experience. New officers were sent for training at Royal Military Academy, Sandhurst which continued until the 1960s and both officers and other ranks were sent to specialist training courses in Britain, India, Pakistan and Malaya. There were no formations and all units were structured to directly function under the Army Headquarters. However temporary field headquarters were to be formed at the time requirement arose.[15]

Due to a lack of any major external threats the growth of the army was slow, and the primary duties of the army quickly moved towards internal security by the mid 1950s, the same time as the first Ceylonese Commander Major General Anton Muttukumaru took command of the army. The first internal security operation of the Ceylon Army began in 1952, code named *Operation Monty* to counter the influx of illegal South Indian immigrants brought in by smugglers on the north-western cost, in support of Royal Ceylon Navy coastal patrols and police operations. This was expanded and renamed as Task Force Anti-Illicit Immigration (TaFII) in 1963 and continued up to 1981 when it was disbanded. The Army was mobilized to help the police to restore peace under provincial emergency regulations during the 1953 hartal, the 1956 Gal Oya Valley riots and in 1958 it was deployed for the first time under emergency regulations throughout the island during the 1958 Riots.[16]

During the 1950s and 1960s the army was called apron to carry to essential services when the workers went on strike which were organized by the left wing parties and trade unions for various reasons, the most notable was the 1961 Colombo Port strike, during which ships threatened to bypass Colombo port and the country almost staved. To counter these common strikes several units were formed, who were employed in development work when there were no strikes.[16]

In 1962 several volunteer officers attempted a military coup, which was stopped hours before it was launched. This attempted coup effect the military to a great extent, since the government mistrusted the military, it reduced the size and growth of the army, especially the volunteer force, with several units being disbanded. In May 1972, when Ceylon was proclaimed a republic and changed its name to from the Dominion of Ceylon to the Republic of Sri Lanka, all Army units were renamed accordingly.[17]

1970–Present

After successfully defeating the insurgency led by the Janatha Vimukthi Peramuna (JVP) in 1971, the army was confronted with a new conflict, this time with the Liberation Tigers of Tamil Eelam (LTTE) and other Tamil militant groups. The war escalated to the point where India was asked to intervene as a peacekeeping force. This was later seen as a tactical error, as the Indian Peace Keeping Force united nationalist elements such as the JVP to politically support the LTTE in their call to evict the IPKF. The war with the LTTE was halted following the signing of a ceasefire agreement in 2002 with the help of international mediation. However, renewed violence broke out in December 2005 and following the collapse of peace talks, the Army has been involved in the heavy fighting that has resumed in the north and east of the country. Since 1980 the army has undertaken many operations against the LTTE rebels. The major operations conducted by the army eventually led to the recapture of Jaffna and other rebel strongholds. In 19 May 2009 Sri Lankan army declare the victory of war as they found the dead body of LTTE leader Velupillai Prabhakaran. This marked the end of the war, with the LTTE ceasing to exist in Sri Lanka as a result of prolonged military offensives conducted by Sri Lanka army.[18]

Major combat operations

- First JVP Insurrection (1971–1972)
- Eelam War I (1976–1987)
  - Operation Liberation
- Second JVP Insurrection (1987–1990)
- Eelam War II (1990–1995)
  - Operation Sea Breeze
  - Operation Thrividha Balaya
  - Operation Balavegaya I, II
- Eelam War III (1995–2002)
  - Operation Riviresa
  - Operation Jayasikuru
  - Operation Rivibala
  - Operation Ranagosa
  - Operation Rivikirana
  - Operation Kinihira I, II, III/IV, V/VI, VII, VIII ,IX
- Eelam War IV (2006–2009)
  - Eastern Theater
    - Operations in Thoppigala
  - Northern Theater
    - Battle of the Forward Defence Lines

Peacekeeping

The Sri Lanka Army has taken part in two peacekeeping missions with United Nations over the course of its history. First assignment was in the Congo (ONUC) (1960–1963). Most recently, following the signing of a ceasefire agreement was signed between the government and the LTTE in 2002, Sri Lankan forces were invited by the United Nations to be part of the UN peacekeeping force in Haiti.[19] The Sri Lanka Light Infantry was selected to be part of the peacekeeping operations, and training programs on counter revolutionary warfare were conducted for the troops in Kukuleganga. Following the successful completion of the training, a battalion under the command of Col. K.A.D.A. Karunasekara left for Haiti on October 22, 2004. In the process of the peacekeeping operations, 2 soldiers were killed in a raid in Petit-Goave.[20] After over 6 months of service, the first contingent of the peacekeeping force returned to Sri Lanka on May 17, 2005.[19] In December 2007, 7th rotation of the Sri Lankan contingent had been deployed with a force of 991 officers and other ranks, many of the those deployed have been awarded the United Nations Medal for their services.[21]

Peacekeeping Sex Scandal

In November 2007, 114[22] members of the 950 member Sri Lankan Army peacekeeping mission in , were accused of sexual misconduct and abuse.[23] [24] 108 members, including 3 officers of the 950-member-strong Sri Lanka peacekeeping contingent is being sent back after being implicated in alleged misconduct and sexual abuse.[25] UN spokeswoman Michele Montas said: "The United Nations and the Sri Lankan government deeply regret any sexual exploitation and abuse that has occurred."[24] The Sri Lankan Officials claim that there is little tangible evidence on this case.[22] After inquiry into the case the UN Office of Internal Oversight Services (OIOS) has concluded 'acts of sexual exploitation and abuse (against children) were frequent and occurred usually at night, and at virtually every location where the contingent personnel were deployed.' The OIOS is assisting in the pending legal proceedings initiated by the Sri Lankan Government and has said charges should include rape "because it involves children under

18 years of age".[26]

Current deployments

As of present, the bulk of the Sri Lankan Army is deployed for domestic defensive and combat operations, while a sizable foreign deployment is maintained.

Domestic

Due to the Sri Lankan Civil War the army has been on a constant mobilized (including reservist) state since the 1980s (except for a brief period from 2002–2005). The majority of the army as been deployed in the North and Eastern provinces of the country, which includes 14 Divisions coming under six operational headquarters and 2 independent Divisions and several independent Brigades. The army is also based in other parts of the island for internal security including a Division for the defense of the capital.

Foreign

The Sri Lanka Army currently participates in several major overseas deployments:

- Haiti - an infantry battalion with support personal totaling around 1000 personal in Haiti as part of the United Nations Stabilization Mission in Haiti since 2004.[27]
- Chad - a contingent of engineers joined the United Nations Mission in the Central African Republic and Chad on May 25, 2010.[28] [29]
- Lebanon - a mechanized infantry company with combat support personal in the United Nations Interim Force in Lebanon since November 2010.[30]

Organization Structure

The professional head of the army is the Commander of the Army, at present Lieutenant General Jagath Jayasuriya. He is assisted by the Chief of Staff of the Army, currently Major General Daya Ratnayaka The Commandant of the Volunteer Force is head of the Army Volunteer Force and is responsible for the administration and recruitment of all reserve units and personal. The Army Headquarters, situated in Colombo is the main administrative and the operational headquarters of the Sri Lanka Army.

Administrative

The Army Headquarters is divided into a number of branches, namely the General Staff (GS) branch responsible for coordination of operations and training and the Adjutant General's (AGs) branch responsible for personal administration, welfare, medical services and rehabilitation. The Quarter Master General's (QMGs) branch is responsible for feeding, transport, movement and construction and maintenance. The Master General of Ordnance's (MGOs) branch is responsible for procurement and maintenance of vehicles and special equipment.[31] The Military Secretary's Branch is responsible for handling all matters pertaining to officers such as promotions, postings and discipline. Each branch is headed by an officer in the rank of Major General who is directly responsible to the Commander of the Army for the smooth functioning of the Branch. Under each Branch, there are several Directorates, each headed by a Brigadier.[31]

The Headquarters of field formations each have its own staff. For instance a divisional headquarters is divided into a GS branch as an AQ branch, each headed by a Colonel and is responsible for operations & training and administration & logistics respectively. Similarly, a Brigade Major and Major AQ is responsible for operations and administration in a brigade.[31]

Like the Indian Army, the Sri Lanka Army has largely retained the British-style regimental system that it inherited upon independence. The individual regiments (such as the Sri Lanka Light Infantry and the Sinha Regiment) operate independently and recruit their own members. Officers tend to remain in a single battalion throughout their careers. The infantry battalion, the basic unit of organization in field operations, includes five companies of four platoons each. Typical platoon have three squads (sections) of ten personnel each. In addition to the basic infantry forces, a commando regiment was also established in 1986. Support for the infantry is provided by an armoured regiment, five reconnaissance regiments, three mechanized infantry regiments, five field artillery regiments, a rocket artillery regiment, three commando regiments , three special forces regiments, six field engineering regiments, five signals battalions, a medical corps, and a variety of logistics units.[32]

Regiments & Corps

[14]

Name	Headquarters	Subunits
Armoured Corps	Rock House Army Camp, Colombo	Eight regular regiments and a volunteer regiment.
Artillery	Panagoda Cantonment, Panagoda	Seven regular regiments and two volunteer regiments.
Engineers	Army Headquarters, Colombo	Six regular regiments and one volunteer regiment.
Signals Corps	Panagoda Cantonment, Panagoda	Seven regular regiments and one volunteer regiment with Signals workshop and Infonation Technologe Uint (11 SLSC)
Light Infantry	Panagoda Cantonment, Panagoda	Ten regular battalions, five volunteer battalions.
Sinha Regiment	Ambepussa Camp, Ambepussa	Seven regular battalions, five volunteer battalions and a head quarter battalion.
Gemunu Watch	Kuruwitha Army camp, Ratnapura	Nine regular units, four volunteer units.
Gajaba Regiment	Saliyapura Camp, Anuradhapura	Twelve regular battalions and five volunteer battalions.
Vijayabahu Infantry Regiment	Boyagane Camp, Kurunegala	Eight regular battalions and four volunteer battalions.
Mechanized Infantry Regiment	N/A	Three regular battalions.
President's Guard	N/A	N/A
Commando Regiment	Ganemulla, Gampaha	Four regular regiments.
Special Forces Regiment	Seeduwa, Gampaha	Three regular regiments.
Military Intelligence Corps	Colombo	Two regular battalions.
Engineer Services Regiment	Panagoda Cantonment, Panagoda	Three regular regiments and a volunteer regiment.
Service Corps	Panagoda Cantonment, Panagoda	Four regular units and one volunteer unit.
Medical Corps	Panagoda Cantonment, Panagoda	Two regular units and one volunteer unit.
Ordnance Corps	Panagoda Cantonment, Panagoda	Three regular ordnance battalions and one volunteer ordnance battalion.
Electrical and Mechanical Engineers	Slave Island, Colombo	Three regular regiments and one volunteer regiment.
Corps of Military Police	Polhengoda, Colombo	Three regular regiments.

General Service Corps	Panagoda Cantonment, Panagoda	Three regular unit and a volunteer unit.(With 3 & 4 SLAGSC (Pay & Recored)
Women's Corps	Regtl Centre, Borella	One regular unit and 4 volunteers unit.
Rifle Corps	Army Headquarters, Colombo	Two volunteer battalions.
Pioneer Corps	Headquarters, Battharamulla, Pelawattha.	One volunteer unit.
National Guard	Kurunegala	32 volunteer battalions.

Operational Command

Organized and controlled by the Army General Staff at Army HQ, various formations are raised from time to time to suit various security requirements and operation in the country and overs seas. The Army at present has deployed 12 Divisions, 7 task forces and several independent brigades. Except for the 11 Division based at the Panagoda Cantonment which is responsible for the maintenance of capability for the defence of the capital, all other divisions, task forces and brigades are deployed for operations in the Northern and Eastern provinces of Sri Lanka, coming under six regional commands known as *Security Forces Headquarters*, which are the Security Forces Headquarters Jaffna (SFHQ-J), Wanni (SFHQ-W), East (SFHQ-E), Kilinochchi (SFHQ-KLN), Mullaittivu (SFHQ-MLT) & South (SFHQ-S)

Each SFHQ and most divisions are commanded by a General Officer Commanding in the rank of Major General. A SFHQ has several divisions under its command and each division is further divided into brigades. Each brigade is commanded by an officer in the rank of Brigadier and has a number of Infantry battalions, support arms (Artillery, Engineers and Signals) and support services (Service Corps, Engineering Services, Ordnance Corps, Electrical and Mechanical Engineers) under assigned to it. There are also several independent brigade (Air Mobile Brigade, Armored Brigade, etc.)

In other parts of the country, there are Area and Sub-Area Headquarters. Armour, Artillery, Engineers and Signals Units are grouped under Brigade Headquarters of their own arm; Armored Brigade, Artillery Brigade and so on.

Force formations

[33]

Security Forces Headquarters - Jaffna (SFHQ-J)

- 51 Division, based in Jaffna
- 52 Division, based in the Jaffna Peninsula
- 55 Division, based in Elephant Pass Military Base, Jaffna Peninsula [33]

Security Forces Headquarters - Wanni (SFHQ-W) [33]

- 56 Division, operating in the Vavuniya District
- 61 Division, operating in the Vavuniya District
- 21 Division
- Area Headquarters Mannar, Mannar

Security Forces Headquarters - East (SFHQ-E)

- 22 Division, based in Trincomalee[34]
- 23 Division, based in Poonani, Batticaloa District [35]

Security Forces Headquarters - Kilinochchi (SFHQ-KLN)[33] [36]

- 57 Division, operating in the Kilinochchi District
- 66 Division, operating in the Kilinochchi District
- 68 Division, operating in the Kilinochchi District

- Task Force 3, operating in the Kilinochchi District
- Task Force 7, operating in the Kilinochchi District

Security Forces Headquarters - Mullaittivu (SFHQ-MLT) [33]

- 59 Division, operating in the Mullaittivu District
- 64 Division, operating in the Mullaittivu District
- 65 Division, Thunukkai, Mullaittivu District
- Task Force 2, operating in the Mullaittivu District

Security Forces Headquarters - South (SFHQ-S)[37] [38]

- Operation Command Colombo, Colombo
- 11 Division, based at the Panagoda Cantonment, Western Province
  - 111 "Kandy" Brigade
- Area Headquarters Hambantota, Hambantota
- Area Headquarters Galle, Galle
- Area Headquarters Diyatalawa, Diyatalawa
- Sub Area Headquarters Ratnapura, Ratnapura
- Sub Area Headquarters Kurunegala, Kurunegala

Independent Divisions

- 53 Division, based at Mankulam [33]
- 58 Division, based at Paranthan (formally referred to as the Task Force 1) [33]

Independent Brigades

- Air Mobile Brigade
- Mechanized Infantry Brigade
- Armored Brigade
- Special Forces Brigade
- Commando Brigade
- Artillery Brigade
- Engineer Brigade
- Signals Brigade

Disbanded

- 54 Division, based at Elephant Pass[39]
- 2 Division
- 3 Division

Training

General Sir John Kotelawala Defence University (KDU) formed in 1981 and situated in Ratmalana, fourteen kilometers south of Colombo, is Sri Lanka's only university specializing in defense studies. Each year, approximately fifty cadets from all three services are admitted to the university (aged 18–22) to participate in a three-year program of academic work and basic training.[31] [32]

Junior field officers of the army and their counterparts in the Navy and Air Force are given advanced training and education at the **Defence Services Command and Staff College** (DSCSC) at Batalanda, Makola which was established in 1997 as the *Army Command and Staff College.*

Basic officer training is carried out at the **Sri Lanka Military Academy** (SLMA) (formally the **Army Training Center**) situated in Diyatalawa, in the Badulla District. The officer cadets graduating from the academy are commissioned as officers in the regular and volunteer forces. The course for officer cadets runs for ninety weeks and includes training in tactics and administration which helps prepare the cadets to take up the positions of platoon commanders. The course consisted of military and academic subjects and also trained the cadets physically. The course helps to promote leadership qualities and the understanding of each one's role as an officer and a servant of the state. Due to the lack of officers within the lower levels, the training process was sped up in the 1980s by developing a short commission course. The cadets were given a training of fifty-six weeks and devoted themselves to continue their careers in the military with the ten years of service for regular army officers and five years of service for volunteer officers.

Training for the new recruits are carried out by the **Army Training School** in Maduru Oya at several locations followed by additional training (both officers and other ranks) at the **Infantry Training Centre** in Minneriya, the **Combat Training School** in Ampara, while non-commissioned officers receive training at the **Non-commissioned Officers Training School** at Kala Oya. All these establishments come under the control of the Directorate of Training, Army Headquarters. Specialist and additional training is given by specialist training schools, regimental training centers and individual field units.[31] [32]

As the armed forces of Sri Lanka have a limited indigenous training facilities, especially in advanced roles, they have depended greatly on military training provided by foreign countries. The United Kingdom played a major role in the early years following independence and have continued to be an important source of military expertise to the Sri Lankan military. Other sources include Pakistan, Australia, Malaysia, and the United States. Additionally, in an agreement reached in 1984, Israeli security personnel (reportedly from Shin Bet, the Israeli counterespionage and internal security organization) trained army officers in counterinsurgency techniques.[32]

The Sri Lankan Army has also provided special training to the United States Army on their request as well as many other countries in military education regarding civilian rescue, jungle combat, and guerilla warfare etc.[8]

Training Centers[40]

1. Sri Lanka Military Academy (SLMA)
2. Army Training School (ATS)
3. Infantry Training Centre (ITC)
4. Combat Training School (CTS)
5. Army Physical Education Centre (APEC)
6. Volunteer Force Training School (VFTS)
7. Marksman Sniper Training School (MSTS)
8. Non-Commission Officer Training School (NCOTS)
9. Language Training School (LTS)
10. Institute of Peacekeeping Support Operations Training Sri Lanka (IPSOT-SL)

Regimental Training Centers[40]

1. Armoured Cops Training Centre
2. School Of Artillery
3. Sri Lanka School of Military Engineering
4. School Of Mechanical Engineers
5. School Of Signals
6. Commando Regiment Training School
7. Engineer Service School
8. Sri Lanka Army Service Corps. School
9. Sri Lanka Army Military School Of Nursing
10. Sri Lanka Army Ordnance School
11. Sri Lanka Electrical And Mechnaical Engineers School
12. Sri Lanka Corps of Military Police School
13. SriLanka Army General Service Corps. School

Personnel

The Sri Lanka Army presently stands at 200,000 strong[2] including 2,960 women plus and 58,000 reservists.[41]

In late 1987, the army had a total estimated strength of up to 40,000 troops, about evenly divided between regular army personnel and reservists on active duty. The approximately 20,000 regular army troops represented a significant increase over the 1983 strength of only 12,000. Aggressive recruitment campaigns following the 1983 riots raised this number to 16,000 by early 1985.[32] By 2007, the Army had expanded to over 120,000 personnel.[3]

Since the Sri Lankan armed forces are all volunteer services, all personal in the Sri Lanka Army have volunteered as regular personnel or reservists. This should not be confused with the traditional term *volunteers* used for reservists or reservist units. Recruitment of the personal are carried island wide with a restrictions in the norther and eastern provinces during the civil war in those areas. The Rifle Corps is the only territorial unit that carries out recruitment from a specific area. In June 2009, Sri Lanka announced plans to create a "Tamil regiment" to promote integration in the army[42]

Parama Weera Vibhushanaya recipients

The Parama Weera Vibhushanaya is the highest award for valour awarded in the Sri Lankan armed forces. Army recipients include;

- Colonel A.F. Lafir †
- Major G. S. Jayanath †
- Captain Saliya Upul Aladeniya †
- Second Lieutenant K.W.T. Nissanka †
- Warrant Officer 2nd Class Pasan Gunasekera †
- Corporal Gamini Kularatne †
- Lance Corporal W. I. M. Seneviratne †

Notable fallen members

Over 23,790 Sri Lankan armed forces personnel were killed since begin of the civil war in 1981 to its end in 2009, this includes 11 general officers killed in active duty or assassinated.[43] 659 service personnel were killed due to the second JVP insurrection from 1987 to 1990. 53 service personnel were killed and 323 were wounded in the first JVP insurrection from 1971 to 1972.[44] Notable fallen members includes;

- Lt. General Denzil Kobbekaduwa † - One of the greatest generals in modern Sri Lanka and Overall Operations Commander, Northern Sector.[45]
- Lt. General Parami Kulatunga † - Former Deputy Chief of Staff of the Army.[45]
- Lt. General Nalin Angammana † - Former GOC 3 Division.[46]
- Maj. General Vijaya Wimalaratne † - Jaffna Brigade Commander & One of the greatest generals in modern Sri Lanka.[45]
- Maj. General Percy Fernando † - Former deputy GOC 54 Division.[46]
- Maj. General Larry Wijeratne † - Former brigade commander, 51-4 Brigade.[46]
- Maj. General Lakshman 'Lucky' Wijayaratne † - Former brigade commander, 22 Brigade.[46]
- Maj. General Susantha Mendis † - Former brigade commander, 51-2 Brigade.[46]
- Maj. General Ananda Hamangoda † - Former brigade commander, 51-2 Brigade.[47]
- Brigadier Bhathiya Jayatilleka † - former Brigade commander, 54-1 brigade
- Brigadier Rohitha Neil Akmeemana † - former Brigade commander, Elephant pass.[46]
- Colonel Tuan Nizam Muthaliff - Former commanding officer 1st Battalion Military Intelligence Corps.[48]

Women in the Sri Lanka Army

The Sri Lanka Army Women's Corps (SLAWC) was formed on September 1, 1979 as an unarmed, noncombatant support unit. Set up with the assistance of the Women's Royal Army Corps, it was identical in structure to its parent organization, and its first generation of officer cadets was trained in Britain. Candidates were required to be between eighteen and twenty years old and to have passed the General Common Entrance (Ordinary level) examinations, while the Officer candidates must have passed the Advanced Level. Enlistment entailed a five-year service commitment (the same as for men), and recruits were not allowed to marry during this period. In the sixteen-week training course at the Army Training Center at the Diyatalawa Sri Lanka Military Academy, cadets were put through a program of drill and physical training similar to the men's program, with the exception of weapons and battle craft training. Female recruits were paid according to the same scale as the men, but were limited to service in nursing, communications, and clerical work. In late 1987, the first class of women graduates from the Viyanini Army Training Center were certified to serve as army instructors. But, from late 1987 - after hostilities began, the first batch of women graduates from the British Army's Women's Corp Center certified to serve as Army Instructors.[49]

Up to now, women officers have proved their ability and serve in varied specialized fields in the Service as control tower operators, electronic warfare technicians, radio material teletypists, automotive mechanics, aviation supply personnel, cryptographers, doctors, combat medic, lawyers, engineers and even aerial photographers.[50]

To meet the operational requirements in the field areas, the 2nd Volunteer Battalion of the Women's Corps was also raised. A few officers from the regular counter part were attached to this unit to organize the command structure. They are currently employed in active combat duties in the northern and eastern parts of the island.

Many officers commencing with Lieutenant Colonel A.W. Thambiraja were appointed to command this unit from time to time. The first women's corps officer to command the unit was Lieutenant Colonel Kumudini Weerasekara in 1992 and as of 2007 there were three female officers of the rank of Major General. At present there is one regular regiment and four volunteer regiments in the Women's Corps.[51]

Complaints over survivors' benefits

The Organisation for Disabled Soldiers has complained that compensation paid for war victims is inadequate. In August 2007, a spokesman for the organisation pointed out that the compensation paid to the families of soldiers killed during the war has remained flat for 23 years at Rs. 150,000, which has dwindled in value to the equivalent of US$1500. This amount is the same for all ranks, including generals. In contrast, however, the families of politicians are much more highly compensated; the family of the late Minister Mr. M. H. M. Ashroff was awarded Rs. 5 million (US$50,000) following his death in a helicopter crash.

Equipment

In the 1980s and 1990s, the army expanded its range of weapons from the original stock of World War II-era British Lee Enfield rifles, Sten Submachine guns, Vickers machine guns, Bren machine guns, 6-inch coastal guns, Daimler Armoured Cars, Bren Gun Carriers,[52] 40 mm anti-aircraft guns, 3.7 inch heavy anti-aircraft guns and 4.2-inch heavy mortars as well as post war Alvis Saladins, Alvis Saracen, Ferrets and Shorland S55s. New sources of weaponry in the mid-to-late 1970s included the Soviet Union, Yugoslavia, and China - countries with which the leftist Bandaranaike government had close ties. China continued to be an important source of arms well into the 1990s.[32]

To meet the threat posed by predominantly the LTTE, Army purchased modern military hardware including 50-caliber heavy machine guns, rocket-propelled grenade (RPG) launchers, Night Vision Devices, 106 mm recoilless rifles, 60 mm and 81 mm mortars, 40 mm grenade launchers and some sniper rifles. Refurbished armored personnel carriers were added to the 'A' vehicle fleet of the 1st Reece Regiment, Sri Lanka Armoured Corps. These APCs enabled the Armoured Corps to have their own assault troops to provide close contact protection to their Alvis Saladin and Ferret Scout Cars which were vulnerable to anti-tank weapons. The capability of the Sri Lanka Artillery was enhanced with the introduction of Ordnance QF 25 pounders.[32] [53] Chinese-made 122 mm, 130 mm and 152 mm howitzers were introduced to the Sri Lankan Army in 1995 and 1998 whilst 122 mm Multi Barrel Rocket Launchers (MBRL), were first used in 2000 by the Sri Lanka Army.[54]

Though the weapons were obsolete at the time of purchase, the Government security forces found them to be successful in combat. Land mines proved to be the most lethal threat to the security forces, as many mines have been deployed against government forces by the LTTE in the Northern and Eastern Provinces. These mines were deployed with normalcy against government trucks and buses, with a high rate of casualty. These land mines weighed approximately 50 – 100 kg, against which no armoured vehicle that the SLA possessed was able to withstand the blast effect. Therefore the South African made Buffel was introduced to service in 1985 to reduce damage due to land mines. By 1987 the Army's own Unicorn was also developed to a level so that they too matched the capabilities of the Buffels from South Africa, this was followed up by the newer Unibuffel class.[55] Both the Unicorn and the Unibuffel are assembled by the Sri Lanka Electrical & Mechanical Engineers.[32] [53]

In recent years, Sri Lanka has become increasingly reliant on China for weapons.[56] This is due to most European nations and the United States Governments passing regulations about the selling of weaponry to nations which are suffering from internal conflict.[57] However in light of recent attacks by the LTTE, the United States has expressed its intent to maintain military training assistance and possibly increase it should the violence continue.

China has no such regulations upon their arms producers, and some see the sales as an attempt to gain political influence with strategically important Sri Lanka.[58] Sri Lanka also continues to receive a variety of weapons from Britain, Pakistan, Israel and other former suppliers.[56] [59]

Armour

Type	Origin	Quantity	Notes
	Main Battle Tanks		
T-55/T-55AM2[55]	Soviet Union	140+	
Type 69[60]	China	20+	
Type 59	China	80+	
	Light Tanks		
Type 63	China	N/A	Amphibious
	Infantry fighting vehicles		
BMP-3[61]	Soviet Union	45	
BMP-2	Soviet Union	240	
BMP-1	Soviet Union	13	
	Armoured personnel carriers		
Norinco Type 89 (YW534)[62]	China	N/A	Tracked
Type 85 (YW531H)	China	N/A	Tracked, Amphibious
Type 63 (YW531)[63]	China	N/A	Tracked
BTR-80	Soviet Union	45	wheeled
Type 92 (WZ551)	China	200	wheeled
Unibuffel	Sri Lanka	53+	Locally manufactured, Mine-protected APC
Unicorn	Sri Lanka	60+	Locally manufactured, Mine-protected APC
Buffel	South Africa	N/A	Mine-protected APC
BTR-152	Soviet Union	N/A	wheeled, (retired)
Hotspur Hussar[31]	United Kingdom	N/A	wheeled, (retired)
Shorland S55	United Kingdom	N/A	wheeled, (retired)
Hotspur[31]	United Kingdom	N/A	wheeled, (retired)
	Reconnaissance vehicles		
Alvis Saladin	United Kingdom	N/A	wheeled (retired)
Alvis Saracen	United Kingdom	N/A	wheeled, (retired)
Daimler Ferret	United Kingdom	N/A	wheeled, (retired)
	Engineering Support Vehicles		

VT-55[64]	Soviet Union	16	Armoured recovery vehicle
MT-55A[65]	Soviet Union	8	Armoured vehicle-launched bridge

Artillery

Type	Origin	Quantity	Notes
Rocket artillery			
RM-70 Multiple rocket launcher[64]	Czechoslovakia	30+	122.4 mm Multiple rocket launcher
BM-21 Multiple rocket launcher[64]	Soviet Union	5	122.4 mm Multiple rocket launcher
Towed artillery			
Type 66 152 mm gun-howitzer	China	36	152 mm gun-howitzer
Type 59 130mm field gun	China	12	130 mm field gun
Type 60 122mm howitzer	China	N/A	122 mm howitzer
Ordnance QF 25 pounder	United Kingdom	N/A	field guns - Ceremonial Gun Troop
Ordnance QF 75 mm	United Kingdom	N/A	field guns - Ceremonial Gun Troop
Mortars			
Type 86 (W86)	China	55	120 mm towed mortar
Type 84 (W84)	China	N/A	82 mm mortar
Type 89	China	N/A	60 mm light mortar
Weapon Locating Radar			
AN/TPQ-36 Firefinder Radar	United States	N/A	Weapon Locating and Counter-battery Radar
SLC-2 Radar	China	N/A	Weapon Locating and Counter-battery Radar

Infantry weapons

Handguns
Beretta M9 Pistol
Glock 17
Enfield revolver
Assault Rifles
AK-47 Assault rifles
Sa 58 Assault rifles
Type 56 Assault rifles[66]
Type 81 Assault rifles[67]
Heckler & Koch G3 Assault rifles[68]
FN FNC Assault rifles (replaced the FN FAL as far back as in 1981)
M16 Assault rifles[69]

M4 Carbine
SAR-80 Assault rifles[66]
Type 95 Assault rifles
Sub-Machine guns
H&K MP5 Submachine Guns[66]
Uzi Submachine Guns[66]
Sterling submachine guns[66]

Sniper Rifles
Accuracy International L96A1 Sniper Rifles[70]
Heckler & Koch PSG1 Sniper Rifles
Dragunov Sniper Rifles[71]
Zastava M93 Black Arrow[72]
Machine guns
Type 80 GPMG (Chinese version of Russian PKM)[73]
FN Minimi General purpose machine gun[66]
FN MAG General purpose machine gun[66]
Type 56 LMG (Chinese version of Russian RPD)[32]
HK21 Belt-fed light machine gun[74]
Grenade launchers
Milkor MGL Grenade launcher[66]
HK 69 Breech-loading grenade launcher[66]
M203 Grenade launcher[66]
Rocket launchers
RPO-A Shmel man-portable rocket launcher
Type 69 RPG rocket launchers (Chinese version of RPG-7)[32]
Anti-tank missiles
Baktar-Shikan Anti-tank guided missile

See also

- Awards and decorations of the military of Sri Lanka
- Sri Lanka Military Academy
- Sri Lankan Civil War
- [[Rehabilitation Directorat, Sri Lanka Army

Further reading

- Army, Sri Lanka. (1st Edition - October 1999). *Sri Lanka army: 50 years on, 1949-1999* ISBN 978-955-8089-02-6

References

[1] http://en.wikipedia.org/wiki/Sri_Lanka_Light_Infantry

[2] http://www.sundayobserver.lk/2010/07/25/sec04.asp

[3] "Country Reports on Human Rights Practices - 2000" (http://www.state.gov/g/drl/rls/hrrpt/2000/sa/index.cfm?docid=704). U.S. State Department. February 23, 2001. . Retrieved 2007-03-23.

[4] Army now 150,000 strong recruits more, readies for big battles on the Vanni front. The Island (http://www.island.lk/2008/02/11/news2.html)

[5] "Deadly arsenals dot Sri Lanka" (http://www.atimes.com/atimes/South_Asia/HH05Df02.html). *Richard M Bennett* (Asia Times Online). 5 August 2006. .

[6] Army, Navy get new commanders (http://www.island.lk/2009/07/13/news20.html)

[7] "Commander-in-Chief Completes One Year in Office" (http://web.archive.org/web/20070928111958/http://www.rnhit.com/mcns/articles/commander-in-chief.html?PHPSESSID=8a9729bc19d560d81319bc8bee3f380a). Media Center for National Security. 2007-03-08. Archived from the original (http://www.rnhit.com/mcns/articles/commander-in-chief.html?PHPSESSID=8a9729bc19d560d81319bc8bee3f380a) on 2007-09-28. . Retrieved 2007-04-20.

[8] http://www.colombopage.com/archive_10A/May15_1273933283JR.php

[9] "Sri Lanka Army History, The Ancient Time" (http://web.archive.org/web/20070202064404/http://www.army.lk/history2.php). *Sri Lanka Army*. Archived from the original (http://www.army.lk/history2.php) on 2007-02-02. . Retrieved 2007-02-04.

[10] Army, Sri Lanka (October 1999). "Chapter 1". *Sri Lanka Army*, "50 YEARS ON" - 1949-1999 (1st Edition ed.). Colombo: Sri Lanka Army. pp. 1–3. ISBN 978-955-8089-02-6.

[11] "The Portuguese in Sri Lanka (1505-1658)" (http://www.lankalibrary.com/geo/portu2.htm). *WWW Virtual Library - Sri Lanka*. . Retrieved 2004-06-30.

[12] "Indias.com, Sri Lankan Army" (http://www.indias.com/php/wiki-Sri_Lankan_Army). *Indias*. . Retrieved 2006-02-04.

[13] "History of the Sri Lanka Light Infantry Regiment, Sri Lanka Army" (http://web.archive.org/web/20060708103051/http://www.army.lk/lightinfan.php). *Sri Lanka Army*. Archived from the original (http://www.army.lk/lightinfan.php) on 2006-07-08. . Retrieved 2006-02-04.

[14] "Establishment, Sri Lanka Army" (http://web.archive.org/web/20060326024640/http://www.army.lk/org.php). *Sri Lanka Army*. Archived from the original (http://army.lk/org.php) on 2006-03-26. . Retrieved 2006-02-04.

[15] Sergei de Silva-Ranasinghe looks back at the early days of the Sri Lanka Army (http://www.sundaytimes.lk/051009/plus/1.html)

[16] An evolving army and its role through time, Sergei de Silva- Ranasinghe's article on the early days of the Sri Lanka Army (http://www.sundaytimes.lk/051016/plus/4.html)

[17] "Sri Lanka Army Marks 50 Years" (http://www.washingtonpost.com/wp-srv/aponline/19991010/aponline085414_000.htm). Washingtonpost, AP News. October 10, 1999. .

[18] Sri Lanka's Tamil Tigers admit defeat (http://www.csmonitor.com/2009/0517/p90s01-wosc.html) Christian Science Monitor - May 17, 2009

[19] "Sri Lanka Light Infantry UN Peacekeeping Operations" (http://web.archive.org/web/20070204125105/http://www.army.lk/lightinfan.php). *Sri Lanka Army*. Archived from the original (http://www.army.lk/lightinfan.php) on 2007-02-04. . Retrieved 2007-03-16.

[20] "2 U.N. peacekeepers killed in Haiti fighting" (http://www.signonsandiego.com/uniontrib/20050321/news_1n21world.html). San Diego Union Tribune. March 21, 2005. .

[21] Sri Lankan troops in Haiti honoured (http://www.dailynews.lk/2008/04/02/news13.asp)

[22] "Haiti: Over 100 Sri Lankan blue helmets repatriated on disciplinary grounds – UN" (http://www.un.org/apps/news/ticker/tickerstory.asp?NewsID=24514). *United Nations*. 2007-11-02. . Retrieved 2007-11-04.

[23] Williams, Carol J. (2007-12-15). "U.N. confronts another sex scandal" (http://articles.latimes.com/2007/dec/15/world/fg-haitisex15). *Los Angeles Times*. . Retrieved 2008-11-07.

[24] "Sri Lanka to probe UN sex claims" (http://news.bbc.co.uk/2/hi/south_asia/7076284.stm). *BBC*. 2007-11-03. . Retrieved 2007-11-05.

[25] Reddy, B. Muralidhar (2007-11-05). "Part of Sri Lankan contingent in Haiti to be sent back" (http://www.hindu.com/2007/11/04/stories/2007110457201800.htm). *The Hindu*. . Retrieved 2007-11-05.

[26] "UN confirms sex charges" (http://www.sundaytimes.lk/080330/News/news0015.html). *Sundaytimes*. 2008-03-30. . Retrieved 2008-03-30.

[27] Haiti - MINUSTAH - Facts and Figures (http://www.un.org/depts/dpko/missions/minustah/facts.html)

[28] http://www.army.lk/detailed.php?NewsId=2195

[29] http://www.dailymirror.lk/print/index.php/news/front-page-news/11435-army-contingent-to-chad-under-un-flag.html

[30] Lankan troops for Lebanon UN peace keeping mission (http://www.island.lk/index.php?page_cat=article-details&page=article-details&code_title=11739)

[31] "Military Balance, Sri Lanka" (http://web.archive.org/web/20061002155351/http://www.ipcs.org/Military_militaryBalance2.jsp?database=1002&country2=Sri+Lanka). *Institute of Peace and Conflict Studies*. Archived from the original (http://www.ipcs.org/Military_militaryBalance2.jsp?database=1002&country2=Sri Lanka) on 2006-10-02. . Retrieved 2007-03-14.

[32] "Sri Lanka, The Army" (http://www.photius.com/countries/sri_lanka/national_security/sri_lanka_national_security_the_army.html). *Photius Coutsoukis*. . Retrieved 2007-03-14.

[33] Two Security Forces Headquarters established in Mullaitivu, Kilinochchi (http://www.dailynews.lk/2009/06/01/sec02.asp)

[34] Security Forces on a realistic path to achieve the set target, Dailynews (http://www.dailynews.lk/2008/07/14/spo10.asp)

[35] Bitter battles, heavy toll, Tiger strongholds under siege (http://www.sundaytimes.lk/081123/Columns/sitreport.html)

[36] Two New SF Headquarters Become Operational (http://www.army.lk/detailed.php?NewsId=697)

[37] New name for JOH (http://www.sundaytimes.lk/090927/News/news_22.html)

[38] Sixth Security Forces Headquarters to Take Charge of Security in South (http://www.army.lk/detailed.php?NewsId=1287)

[39] Eelam War IV: Tigers Face Defeat (http://www.army.lk/morenewsfet.php?id=5272)

[40] "Training, Sri Lanka Army" (http://web.archive.org/web/20060410062148/http://www.army.lk/train.php). *Sri Lanka Army*. Archived from the original (http://army.lk/train.php) on 2006-04-10. . Retrieved 2006-02-04.

[41] "Sri Lanka" (http://www.nationsencyclopedia.com/Asia-and-Oceania/Sri-Lanka-ARMED-FORCES.html). *Nations Encyclopedia*. . Retrieved 2007-02-04.

[42] 'Tamil unit' for Sri Lanka's army (http://news.bbc.co.uk/2/hi/south_asia/8124836.stm) BBC - June 29, 2009

[43] "Victory's price: 6,200 Sri Lankan troops" (http://news.smh.com.au/breaking-news-world/victorys-price-6200-sri-lankan-troops-20090522-bi4f.html). News.smh.com.au. 22 May 2009. . Retrieved 30 May 2009.

[44] Crushing the revolt (http://sundaytimes.lk/010408/spec.html)

[45] Kanagaraarachchi, Ramani (2007-01-20). "Heroes who made the supreme sacrifice" (http://www.dailynews.lk/2007/01/20/spice01.asp). *Daily News* (Associated Newspapers of Ceylon). . Retrieved 2007-05-25.

[46] Heroes who made the supreme sacrifice (http://www.dailynews.lk/2007/01/20/spice01.asp)

[47] Ratnayake, Brigadier Daya (2005-07-30). "Noble in death as in life" (http://sundaytimes.lk/050703/plus/appreciation.html). *The Sunday Times* (Wijeya Newspapers Ltd). . Retrieved 2007-08-18.

[48] De Silva, Senaka (2005-06-01). "Senior Army officer shot dead" (http://web.archive.org/web/20070930224634/http://www.dailymirror.lk/2005/06/01/front/1.asp). *Daily Mirror* (Wijeya Newspapers Ltd). Archived from the original (http://www.dailymirror.lk/2005/06/01/front/1.asp) on 2007-09-30. . Retrieved 2007-06-03.

[49] "Sri Lankan Army Women's Corps" (http://womenshistory.about.com/library/ency/blwh_sri_lanka_women_military.htm). *About, Inc.*. Retrieved 2007-02-04.

[50] "An officer and a lady: You've come a long way, lass." (http://www.sundayobserver.lk/2006/10/08/zin01.asp). *Sunday Observer*. . Retrieved 2006-10-08.

[51] "Sri Lanka Army Women's Corps, Sri Lanka Army" (http://web.archive.org/web/20070128135530/http://www.army.lk/regiment_regular_19.php). *Sri Lanka Army*. Archived from the original (http://www.army.lk/regiment_regular_19.php) on 2007-01-28. . Retrieved 2007-02-04.

[52] Michael K. Cecil - Sri Lanka's Military: The Search For A Mission, (http://www.bdblodgett.com/srilanka.pdf).

[53] Army, Sri Lanka (October 1999). "Chapter 1". *Sri Lanka Army*, "50 YEARS ON" - 1949-1999 (1st Edition ed.). Colombo: Sri Lanka Army. p. 393. ISBN 978-955-8089-02-6.

[54] http://www.army.lk/org2.php

[55] "SFrom Unicorn to Unibuffel" (http://www.sundayobserver.lk/2004/04/25/fea22.html). sundayobserver. .

[56] Bedi, Rahul (2007-06-02). "Sri Lanka turns to Pakistan, China for military needs" (http://www.indianmuslims.info/news/2007/jun/02/sri_lanka_turns_pakistan_china_military_needs.html). *IANS* (Urdustan.com Network). . Retrieved 2007-06-02.

[57] "Small Arms and Light Weapons Conference Report" (http://www.globalactionpw.org/un/SALW_PoA_RC_ REPORT.pdf) (PDF). *Judith McDaniel*. . Retrieved 2006-07-23.

[58] "No Nonsense guide to The Arms Trade" (http://www.thirdworldtraveler.com/Weapons/Arms_Trade.html). *Gideon Burrows*. . Retrieved 2005-11-19.

[59] "Naval surveillance is the millstone around LTTE's neck" (http://www.priu.gov.lk/news_update/features/20031017naval_surveillance_millstone_LTTE.htm). *P.K. Balachanddran*. . Retrieved 2003-10-17.

[60] "Type 69 Main Battle Tank" (http://www.globalsecurity.org/military/world/china/type-69.htm). GlobalSecurity.org. July 31, 2005. . Retrieved 2007-03-15.

[61] "Sri Lanka (LTTE)" (http://acd.iiss.org/armedconflict/Mainpages/dsp_ConflictWeapons.asp?ConflictID=174). International Institute For Strategic Studies. March 15, 2007. . Retrieved 2007-03-15.

[62] "ZSD89 Armoured Personnel Carrier" (http://www.sinodefence.com/army/armour/zsd89.asp). sinodefence.com. 2007. . Retrieved 2007-12-02.

[63] [//www.nation.lk/2006/10/15/newsfe1.htm "Type 63"]. nation.lk. //www.nation.lk/2006/10/15/newsfe1.htm.

[64] "United Nations Register of Conventional Arms" (http://disarmament.un.org/library.nsf/c793d171848bac2b85256d7500700384/c7b9a51a4973f6e585256dc100585789/$FILE/sg58.203.pdf) (PDF). *United Nations.* August 21, 2003. . Retrieved 2007-03-16.

[65] Saferworld's research project on arms and security in EU Associate Countries, Czech Republic (http://www.saferworld.org.uk/images/pubdocs/Beast CzechRep.pdf)

[66] Smith, Chris (October 2003) (PDF). *In the Shadow of a Cease-fire: The Impacts of Small Arms Availability and Misuse in Sri Lanka* (http://www.smallarmssurvey.org/files/sas/publications/o_papers_pdf/2003-op11-sri_lanka.pdf). Small Arms Survey. .

[67] "T 81 Assault Rifle" (http://web.archive.org/web/20070214094032/http://www.sinodefence.com/army/small_arms/type81rifle.asp). *SinoDefence.* Archived from the original (http://www.sinodefence.com/army/small_arms/type81rifle.asp) on 2007-02-14. . Retrieved 2007-03-14.

[68] International, Amnesty; International Secretariat, AI Ireland, Omega Foundation (PDF). *Undermining Global Security: the European Union's arms exports* (http://www.smallarmssurvey.org/files/portal/spotlight/ammunition/ammun_pdf/2004_amnesty_international.pdf) (1st Edition ed.). Amnesty International. ISBN 0-86210-356-8. . Retrieved 14 March 2007.

[69] "Profiling the Small Arms Industry" (http://www.worldpolicy.org/projects/arms/reports/smallarms.htm). *Arms Trade Resource Center.* World Policy Institute. November 2000. . Retrieved 2007-03-15.

[70] "Accuracy International L96A1" (http://www.militaryfactory.com/smallarms/detail.asp?smallarms_id=34). www.MilitaryFactory.com. January 16, 2009. . Retrieved 2009-01-20.

[71] "Trans-National Crime and Light Weapons Proliferation: Security Implications for the State" (http://web.archive.org/web/20061207164518/http://www.cpdsindia.org/globalhumansecurity/trans-national.htm). *Tara Kartha.* IDSA. August 15, 2002. Archived from the original (http://www.cpdsindia.org/globalhumansecurity/trans-national.htm) on 2006-12-07. . Retrieved 2007-03-15.

[72] (http://www.sipri.org/research/armaments/transfers/transparency/national_reports/serbia/SER_08.pdf/view)

[73] "Biography of Mikhail Kalashnikov" (http://www.vodkakalashnikov.com/biography.htm). VRQ International, Inc.. January 14, 2004. . Retrieved 2007-03-15.

[74] "If You're Involved in Resolving the" (http://www.ibiblio.org/obl/reg.burma/archives/199901/msg00057.html). ibiblio.org. January 5, 1999. . Retrieved 2007-03-15.

72. http://www.spur.asn.au/extra/Curr10.htm

External links

- Official website (http://www.army.lk)
- Ministry of Defence Sri Lanka (http://www.defence.lk/)
- General Sir John Kotelawala Defence Academy (http://www.kda.lk/)

Panagoda_Cantonment

Panagoda Cantonment	
Panagoda, Western Province	
Type	Cantonment
Built	1953
In use	1953 – present
Garrison	11 Division, Sri Lanka Artillery, Sri Lanka Signals Corps, Sri Lanka Light Infantry

Panagoda Cantonment is a cantonment located in the western Western Province of Sri Lanka. It serves as the regimental headquarters of many regiments of the Sri Lanka Army and is an arsenal. The 11 Division of the Sri Lanka Army is based at the Panagoda cantonment.[1] It also houses one of the main Military Hospitals operated by the Sri Lanka Army Medical Corps.

List of regiment which as its regimental headquarters at the Panagoda Cantonment

- Sri Lanka Light Infantry
- Sri Lanka Artillery
- Sri Lanka Signals Corps
- Engineer Services Regiment
- Sri Lanka Army Service Corps
- Sri Lanka Army Ordnance Corps
- Sri Lanka Army Medical Corps
- Sri Lanka Army General Service Corps

References

[1] Overall operations commander, A'Pura (http://www.dailynews.lk/2007/10/24/sec03.asp)

Artillery

Originally applied to any group of infantry primarily armed with projectile weapons, **artillery** has over time become limited in meaning to refer only to those engines of war that operate by projection of munitions far beyond the range of effect of personal weapons. These engines comprise specialised devices which use some form of stored energy to operate, whether mechanical, chemical, or electromagnetic. Originally designed to breach fortifications, they have evolved from nearly static installations intended to reduce a single obstacle to highly mobile weapons of great flexibility in which now reposes the greater portion of a modern army's offensive capabilities.

French naval piece of the late 19th century

In common speech the word artillery is often used to refer to individual devices, together with their accessories and fittings, although these assemblages are more properly referred to as equipments. By association, artillery may also refer to the arm of service that customarily operates such engines.

Artillery may also refer to a system of applied scientific research relating to the design, manufacture and employment of artillery weapon systems although, in general, the terms ballistics and ordnance are more commonly employed in this sense.

Artillery is by far the deadliest and most effective form of land-based armament; in the Napoleonic Wars, World War I and World War II the vast majority of combat deaths were caused by artillery.[1] In 1944, Joseph Stalin said in a speech that artillery was "the God of War".[1]

Overview

Although not called as such, machines recognizable as artillery have been employed in warfare since antiquity. The first references in the western historical tradition may be those of Hero of Alexandria c. 1st century AD/CE, but these devices were widely employed by the Roman Legions in Republican times well before the Christian era. Through much of their early history artillery was treated as part of the engineering art because the devices were often constructed mostly of local materials whenever needed and not permanently assembled. Until the introduction of gunpowder into western warfare artillery depended upon mechanical energy to operate and this severely limited the range and size of projectiles while also requiring the construction of very large apparatus to store sufficient energy.

French soldiers in the Franco-Prussian War 1870-71.

For much of artillery's history during the Middle Ages and the Early modern period, artillery pieces on land were moved with the assistance of horse teams. During the more recent Modern era and in the Post-Modern period the artillery crew has used wheeled or tracked vehicles as a mode of transportation. Artillery used by naval forces has changed significantly also, with missiles replacing guns in surface warfare.

Over the course of military history, projectiles were manufactured from a wide variety of materials, made in a wide variety of shapes, and used different means of inflicting physical damage and casualties to defeat specific types of targets. The engineering designs of the means of delivery have likewise changed significantly over time, and have become some of the most complex technological application today.

In some armies, the weapon of artillery is the projectile, not the piece that fires it. The process of delivering fire onto the target is called gunnery. The actions involved in operating the piece are collectively called "serving the gun" or "detachment" by the gun crew, constituting either direct or indirect artillery fire. The manner in which artillery units or formations are employed is called artillery support, and may at different periods in history refer to weapons designed to be fired from ground, sea, and even air-based weapons platforms.

Although the term also describes soldiers and sailors with the primary function of using artillery weapons, the individuals who operate them are called gunners whatever their rank, however 'gunner' is the lowest rank in artillery arms. There is no generally recognised generic term for a gun, howitzer, mortar, and so forth: some armies use 'artillery piece', while others use 'gun'. The projectiles fired are typically either 'shot' (if solid) or 'shell' if not. Shell is a widely used generic term for a projectile, which is a component of munitions.

The term 'artillery' is also applied to a combat arm of most military services when used organizationally to describe units and formations of the national armed forces that operate the weapons.

The gunners and their guns are usually grouped in teams called either 'crews' or 'detachments'. Several such crews and teams with other functions are combined into a unit of artillery usually called a battery, although sometimes called a company. Batteries are roughly equivalent to a company in the infantry, and are combined into larger military organizations for administrative and operational purpose.

5-person gun crew firing a US M777 Light Towed Howitzer, War in Afghanistan, 2009.

During military operations the role of field artillery is to provide close support to other arms in combat or to attack targets. The latter role is typically achieved by delivering either high explosive munitions to inflict casualties on the enemy from casing fragments and other debris and blast, or by demolition of enemy positions, equipment and vehicles. Fire may be directed by an artillery observer or called onto map coordinates.

Military doctrine has played a significant influence on the core engineering design considerations of Artillery ordnance through its history, in seeking to achieve a balance between delivered volume of fire with ordnance mobility. However, during the modern period the consideration of protecting the gunners also arose due to the late-19th century introduction of the new generation of infantry weapons using conoidal bullet, better known as the Minié ball, with a range almost as long as that of field artillery.

The gunners' increasing proximity to and participation in direct combat against other combat arms and attacks by aircraft made the introduction of a gun shield necessary. The problems of how to employ a fixed or horse towed gun in mobile warfare necessitated the development of new methods of transporting the artillery into combat. Two distinct forms of artillery developed: the towed gun, which was used primarily to attack or defend a fixed line; and the self-propelled gun, which was designed to accompany a mobile force and provide continuous fire support. These influences have guided the development of artillery ordnance, systems, organisations, and operations until the present, with artillery systems capable of providing support at ranges from as little as 100 m to the intercontinental ranges of ballistic missiles. The only combat in which artillery is unable to take part in is close quarters combat.

Etymology

The word as used in the current context originated in the Middle Ages. One suggestion is that it comes from the Old French *atellier* meaning "to arrange", and *attillement* meaning "equipment".

From the 13th Century an *artillier* referred to a builder of any war equipment, and for the next 250 years the sense of the word "artillery" covered all forms of military weapons. Hence the naming of the Honourable Artillery Company an essentially infantry unit until the 19th Century. Another suggestion is that comes from the Italian *arte de tirare* (art of shooting) coined by one of the first theorists on the use of artillery, Niccolo Tartaglia.

History

Mechanical systems used for throwing ammunition in ancient warfare, also known as "engines of war", like the catapult, onager, trebuchet, and ballista, are also referred to by military historians as artillery.[2]

Roman mechanical artillery (click on picture for detailed descriptions)

High Middle Ages – first gunpowder artillery

The first documented record of artillery with gunpowder propellant used on the battlefield was on January 28, 1132, when General Han Shizhong of the Song Dynasty used escalade and Huochong to capture a city in Fujian.[3] These small, crude weapons diffused into the Middle East (the *madfaa*) and reached Europe in the 13th Century, in a very limited manner.

In Asia, Mongols adopted the Chinese artillery and used it effectively in the great conquest. By late 14th century, Chinese rebels used organized artillery and cavalry to push Mongols out.

As small smoothbore tubes these were initially cast in iron or bronze around a core, with the first drilled bore ordnance recorded in operation near Seville in 1247. They fired lead, iron, or stone balls, sometimes large arrows and on occasions simply handfuls of whatever scrap came to hand. During the Hundred Years' War, these weapons became more common, initially as the bombard and later the cannon. Cannon were always muzzle-loaders. While there were many early attempts at breech-loading designs, a lack of engineering knowledge rendered these even more dangerous to use than muzzle-loaders.

The 15th century

French gunner in the 15th century, a 1904 illustration.

In 1415, the Portuguese invaded the Mediterranean port town of Ceuta. While it is difficult to confirm the use of firearms in the siege of the city, it is known the Portuguese defended it thereafter with firearms, namely *bombardas*, *colebratas*, and *falconetes*. In 1419, Sultan Abu Sa'id led an army to reconquer the fallen city, and Moroccans brought cannons and used them in the assault on Ceuta. Finally, hand-held firearms and riflemen appear in Morocco, in 1437, in an expedition against the people of Tangiers.[5] It is clear these weapons had developed into several different forms, from small guns to large artillery pieces.

The artillery revolution in Europe caught on during the Hundred Years' War and changed the way that battles were fought. In the following year, the English used a gunpowder weapon in a military campaign against the Scottish. However, at this time, the cannons used in battle were very small and not particularly powerful. Cannons were only useful for the defense of a castle, as demonstrated at Breteuil in 1356, when the besieged English used a cannon to destroy an attacking French assault tower. By the end of the 14th Century, cannons were only powerful enough to knock in roofs, and could not penetrate castle walls.

Cannon in the Castle of São Jorge, Lisbon.

However, a major change occurred between 1420–1430, when artillery became much more powerful and could now batter strongholds and fortresses quite efficiently. The English, French, and Burgundians all advanced in military technology, and as a result the traditional advantage that went to the defense in a siege was lost. The cannons during this period were elongated, and the recipe for gunpowder was improved to make it three times as powerful as before.[6] These changes led to the increased power in the artillery weapons of the time.

The Austrian Pumhart von Steyr, the earliest extant supergun.[4]

Joan of Arc encountered gunpowder weaponry several times. When she led the French against the English at the Battle of Tourelles, in 1430, she faced heavy gunpowder fortifications, and yet her troops prevailed in that battle. In addition, she led assaults against the English-held towns of Jargeau, Meung, and Beaugency, all with the support of large artillery units. When she led the assault on Paris, Joan faced stiff artillery fire, especially from the suburb of St. Denis, which ultimately led to her defeat in this battle. In April 1430, she went to battle against the Burgundians, whose support was purchased by the English. At this time, the Burgundians had the strongest and largest gunpowder arsenal among the European powers, and yet the French, under Joan of Arc's leadership, were able to beat back the Burgundians and defend themselves.[7] As a result, most of the battles of the Hundred Years' War that Joan of Arc participated in were fought with gunpowder artillery.

The army of Mehmet the Conqueror, which conquered Constantinople in 1453, included both artillery and foot soldiers armed with gunpowder weapons.[8] The Ottomans brought to the siege sixty-nine guns in fifteen separate batteries and trained them at the walls of the city. The barrage of Ottoman cannon fire lasted forty days, and they are estimated to have fired 19,320 times.[9] Artillery also played a decisive role in the Battle of St. Jakob an der Birs of 1444.

A mortar, Malbork Castle

Early modern period – age of the smoothbores

The Tsar Cannon(caliber 890 mm), cast in 1586 in Moscow. It is the largest bombard in the world

The Dardanelles Gun, cast in 1464 and of the type used by the Ottoman besiegers of Constantinople in 1453. Today it belongs to the British Royal Armouries collection.

Bombards were of value mainly in sieges, a famous Turkish example used at the siege of Constantinople in 1453 weighed 19 tons, took 200 men and sixty oxen to emplace, and could fire just seven times a day. The Fall of Constantinople was perhaps "*the first event of supreme importance whose result was determined by the use of artillery*" when the huge bronze cannons of Mehmed II breached the city's walls, ending the Byzantine Empire, according to Sir Charles Oman.[10]

Bombards developed in Europe were massive smoothbore weapons distinguished by their lack of a field carriage, immobility once emplaced, highly individual design, and noted unreliability (in 1460 James II, King of Scots, was killed when one exploded at the siege of Roxburgh). Their large size precluded the barrels being cast and they were constructed out of metal staves or rods bound together with hoops like a barrel, giving their name to the gun barrel.[11]

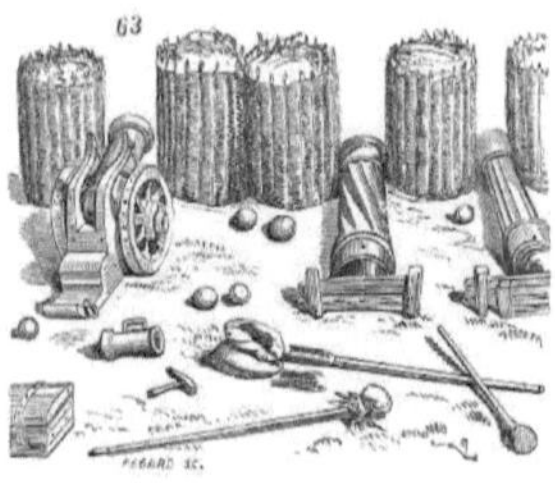

Artillery with Gabion fortification.

Cannons on display at Fort Point.

The use of the word "cannon" marks the introduction in the 15th Century of a dedicated field carriage with axle, trail and animal-drawn limber—this produced mobile field pieces that could move and support an army in action, rather than being found only in siege and static defences. The reduction in the size of the barrel was due to improvements in both iron technology and gunpowder manufacture, while the development of trunnions – projections at the side of the cannon as an integral part of the cast – allowed the barrel to be fixed to a more movable base, and also made raising or lowering the barrel much easier.[11]

The first land-based mobile weapon is usually credited to Jan Žižka, who deployed his oxen-hauled cannon during the Hussite Wars of Bohemia (1418–1424). However cannons were still large and cumbersome. With the rise of musketry in the 16th Century, cannon were largely (though not entirely) displaced from the battlefield—the cannon were too slow and cumbersome to be used and too easily lost to a rapid enemy advance.

Polish multiple gun from 16th-17th Century

The combining of shot and powder into a single unit, a cartridge, occurred in the 1620s with a simple fabric bag, and was quickly adopted by all nations. It speeded loading and made it safer, but unexpelled bag fragments were an additional fouling in the gun barrel and a new tool—a worm—was introduced to remove them. Gustavus Adolphus is identified as the general who made cannon an effective force on the battlefield—pushing the development of much lighter and smaller weapons and deploying them in far greater numbers than previously. The outcome of battles was still determined by the clash of infantry.

Shells, explosive-filled fused projectiles, were also developed in the 17th Century. The development of specialized pieces—shipboard artillery, howitzers and mortars—was also begun in this period. More esoteric designs, like the multi-barrel *ribauldequin* (known as "organ guns"), were also produced.

The 1650 book by Kazimierz Siemienowicz "*Artis Magnae Artilleriae pars prima*" [12] was one of the most important contemporary publications on the subject of artillery. For over two centuries this work was used in Europe as a basic artillery manual.

One of the most significant effects of artillery during this period was however somewhat more indirect – by easily reducing to rubble any medieval-type fortification or city wall (some which had stood since Roman times), it abolished millennia of siege-warfare strategies and styles of fortification building.[11] This led, among other things, to a frenzy of new bastion-style fortifications to be built all over Europe and in its colonies, but also had a strong integrating effect on emerging nation-states, as kings were able to use their newfound artillery superority to force any local dukes or lords to submit to their will, setting the stage for the absolutist kingdoms to come.[11]

Modern era – age of rifled guns

The highly effective German 15 cm sFH 1315cm field howitzers during the First World War

Cannons continued to become smaller and lighter—Frederick II of Prussia deployed the first genuine light artillery during the Seven Years' War—but until the mid-19th Century improvements in metallurgy, chemistry, manufacturing and other sciences did not alter the basic design and operation of a cannon.

"The earliest battlefield use of indirect fire was probably at Paltzig in July 1759: the Russian artillery fired over the tops of trees."[13] Artillery continued to gain prominence in the 18th Century when Jean-Baptiste de Gribeauval, a French artillery engineer introduced the standardization of cannon design. He developed a 6-inch (150 mm) field howitzer whose gun barrel, carriage assembly and ammunition specifications were made uniform for all French cannons. The standardized interchangeable parts of these cannons down to the nuts, bolts and screws made their mass production and repair much easier.

Another major change at this time was the development of a flintlock firing mechanism for the cannons. The old method of firing the cannon involved the use of a linstock or match to light a small quantity of powder charge in a touchhole drilled into the breech. This technique was quite faulty because the ignited powder could easily be extinguished by rain and an excess amount of charge could cause the guns to burst.

The flintlock mechanism on the other hand only needs to be cocked and when its trigger is pulled the flint of the hammer strikes the frizzen throwing sparks into the pan and detonating the charge at the breech. The trigger can be tied to a lanyard and fired from a safe distance. These changes laid down in 1789 proved decisive for Napoleon's conquests. Napoleon, himself a former artillery officer, perfected the tactic of massed artillery batteries unleashed upon a critical point in his enemies' line as prelude to infantry and cavalry assault and, more often than not, victory.

Rifling had been tried on small arms in the 15th Century. The machinery to accurately rifle a cannon barrel did not arrive until the 19th Century. Cavelli, Wahrendorff, and Whitworth all independently produced rifled cannon in the 1840s, but these guns did not see widespread use until the latter stages of the American Civil War, when designs such as the various caliber Rodman guns came to prominence.

From the 1860s artillery was forced into a series of rapid technological and operational changes, accelerating through the 1870s and thereafter. The first effective breech-loader (allowing a higher rate of fire while keeping the detachment behind the gun) was developed in 1855 by Sir William Armstrong, and accepted for British service in 1859. The first cannon to contain all 'modern' features is generally considered to be the French 75 of 1897 with its cased ammunition, effective breech-loading, modern sights, self-contained firing mechanism, and hydro-pneumatic recoil dampening.

After the War of 1870, the Germans became strong advocates of indirect fire. In 1882 a Russian officer, Lieutenant Colonel KG Guk, published *Indirect Fire for Field Artillery* that provided a practical method of using aiming points for indirect fire by describing, "all the essentials of aiming points, crest clearance, and corrections to fire by an observer".[14] A few years later the Richtfläche (lining-plane) sight was invented in Germany and provided a means of indirect laying in azimuth, complementing the clinometers for indirect laying in elevation which already existed. In the next 15 years, the techniques of indirect fire became available for all types of artillery. Indirect fire was the defining characteristic of 20th Century artillery and led to undreamt of changes in the amount of artillery, its tactics, organisation and techniques most of which occurred during World War I. To quote McCamley,

> [By WWII] decades if not centuries of weapons development had settled into maturity on an almost imperceptibly rising plateau; the sciences of ballistics and explosive chemistry had achieved near perfection given the available technology of the age. Arguably the only new developments of note were discarding sabot rounds... and the hollow-charge projectile... both of which were of marginal

significance in the Second World War.

The movement toward precision fires did not begin until late in World War I. "In 1916, the Russians conducted their first registration point shoot: the adjustment of fire onto one target to identify inaccuracy, then switching to another while applying the known correction to achieve accuracy and surprise."[15] Artillery had begun the shift away from massed fires to a greater emphasis on precision fires.

Soviet Katyusha rocket artillery firing in the Battle of Berlin.

According to Jonathan B. A. Bailey, "From the middle of the eighteenth century to the middle of the nineteenth, artillery is judged to have accounted for perhaps 50% of battlefield casualties. In the sixty years preceding 1914, this figure was probably as low as 10 percent. The remaining 90 percent fell to small arms, whose range and accuracy had come to rival those of artillery. ... [By WWI] The British Royal Artillery, at over one million men, grew to be larger than the Royal Navy. Bellamy (1986), pp. 1-7, cites the percentage of casualties caused by artillery in various theaters since 1914: in the First World War, 45 percent of Russian casualties and 58 percent of British casualties on the Western Front; in the Second World War, 75 percent of British casualties in North Africa and 51 percent of Soviet casualties (61 percent in 1945) and 70 percent of German casualties on the Eastern Front; and in the Korean War, 60 percent of US casualties, including those inflicted by mortars."[16]

After the Second World War – age of precision

Modern artillery is most obviously distinguished by its large caliber, firing an explosive shell or rocket, and being of such a size and weight as to require a specialized carriage for firing and transport. However, its most important characteristic is the use of indirect fire, whereby the firing equipment is aimed without seeing the target through its sights. Indirect fire emerged at the beginning of the 20th Century and was greatly enhanced by the development of predicted fire methods in World War I. Indirect fire uses firing data set on the sights, predicted fire methods ensure that these data are accurate and corrects for variations from the standard conditions for muzzle velocity, temperature, wind, and air density.

World War I German Field Gun

Weapons covered by the term 'modern artillery' include "cannon" artillery (such as howitzer, mortar, and field gun) and rocket artillery. Certain smaller-caliber mortars are more properly designated small arms rather than artillery, albeit indirect-fire small arms. This term also came to include coastal artillery which traditionally defended coastal areas against seaborne attack and controlled the passage of ships. With the advent of powered flight at the start of the 20th Century, artillery also included ground-based anti-aircraft batteries.

The term "artillery" has traditionally not been used for projectiles with internal guidance systems, preferring the term "missilery", though some modern artillery units employ surface-to-surface missiles. Advances in terminal guidance systems for small munitions has allowed large-caliber guided projectiles to be developed, blurring this distinction.

Ammunition

One of the most important role of logistics is the supply of munitions as a primary type of artillery consumable, their storage and the provision of fuses, detonators and warheads at the point where artillery troops will assemble the charge, projectile, bomb or shell.

Artillery ammunition can also make use of nuclear warheads, as seen here.

A round of artillery ammunition comprises four components:

- 1: The fuze
- 2: The projectile
- 3: The propellant
- 4: The primer

Fuzes

The normal artillery spelling is "fuze". Fuzes are the devices that trigger explosion of the artillery ammunition charge. Broadly there are four main types:

- impact (including graze and delay)
- mechanical time including airburst
- proximity sensor including airburst
- electronic time including airburst

Most artillery fuzes are nose fuzes. However, base fuzes have been used with armour piercing shells and for squash head (HESH or HEP) anti-tank shells. At least one nuclear shell and its non-nuclear spotting version also used a multi-deck mechanical time fuze fitted into its base.

Impact fuzes were, and in some armies remain, the standard fuze for HE. Their default action is normally 'superquick', some have had a 'graze' action which allows them to penetrate light cover and others have 'delay'. Delay fuzes allow the shell to penetrate the ground before exploding. Armor- or concrete-piercing fuzes are specially hardened. During World War I and later, ricochet fire with delay or graze fuzed HE shells, fired with a flat angle of descent, was used to achieve airburst.

HE shells can be fitted with other fuzes. Airburst fuzes usually have a combined airburst and impact function. However, until the introduction of proximity fuzes, the airburst function was mostly used with cargo munitions—for example shrapnel, illumination, and smoke. The larger calibers of anti-aircraft artillery are almost always used airburst. Airburst fuzes have to have the fuze length (running time) set on them. This is done just before firing using either a wrench or a fuze setter pre-set to the required fuze length.

Early airburst fuzes used igniferous timers which lasted into the second half of the 20th Century. Mechanical time fuzes appeared in the early part of the century. These required a means of powering them. The Thiel mechanism used a spring and escapement (i.e. 'clockwork'), Junghans used centrifugal force and gears, and Dixi used centrifugal force and balls. From about 1980, electronic time fuzes started replacing mechanical ones for use with cargo munitions.

Proximity fuzes have been of two types: photo-electric or radar. The former was not very successful and seems only to have been used with British anti-aircraft artillery 'unrotated projectiles' (rockets) in World War II. Radar proximity fuzes were a big improvement over the mechanical (time) fuzes which they replaced. Mechanical time fuzes required an accurate calculation of their running time, which was affected by non-standard conditions. With HE (requiring a burst 20 to 30 feet (9.1 m) above the ground), if this was very slightly wrong the rounds would either hit the ground or burst too high. Accurate running time was less important with cargo munitions that burst much higher.

The first radar proximity fuzes (codenamed 'VT') were initially used against aircraft in World War II. Their ground use was delayed for fear of the enemy recovering 'blinds' (artillery shells which failed to detonate) and copying the fuze. The first proximity fuzes were designed to detonate about 30 feet (9.1 m) above the ground. These air-bursts are much more lethal against personnel than ground bursts because they deliver a greater proportion of useful fragments and deliver them into terrain where a prone soldier would be protected from ground bursts.

However, proximity fuzes can suffer premature detonation because of the moisture in heavy rain clouds. This led to 'controlled variable time' (CVT) after World War II. These fuzes have a mechanical timer that switched on the radar about 5 seconds before expected impact, they also detonated on impact.

The proximity fuze emerged on the battlefields of Europe in late December 1944. They have become known as the U.S. Artillery's "Christmas present", and were much appreciated when they arrived during the Battle of the Bulge. They were also used to great effect in anti-aircraft projectiles in the Pacific against *kamikaze* as well as in Britain against V-1 flying bombs.[17]

Electronic multi-function fuzes started to appear around 1980. Using solid-state electronics they were relatively cheap and reliable, and became the standard fitted fuze in operational ammunition stocks in some western armies. The early versions were often limited to proximity airburst, albeit with height of burst options, and impact. Some offered a go/no-go functional test through the fuze setter.

Later versions introduced induction fuze setting and testing instead of physically placing a fuze setter on the fuze. The latest, such as Junghan's DM84U provide options giving, superquick, delay, a choice of proximity heights of burst, time and a choice of foliage penetration depths.

A new type of artillery fuze will appear soon. In addition to other functions these offer some course correction capability, not full precision but sufficient to significantly reduce the dispersion of the shells on the ground.

Projectiles

The projectile is the munition or "bullet" fired downrange. This may or may not be an explosive device. Traditionally, projectiles have been classified as "shot" or "shell", the former being solid and the latter having some form of "payload".

Shells can also be divided into three configurations: bursting, base ejection or nose ejection. The latter is sometimes called the shrapnel configuration. The most modern is base ejection, which was introduced in World War I. Both base and nose ejection are almost always used with airburst fuzes. Bursting shells use various types of fuze depending on the nature of the payload and the tactical need at the time.

Payloads have included:

- Bursting: high-explosive, white phosphorus ("Willie Pete" or "Wilson Picket"), coloured marker, chemical, nuclear devices; high explosive anti-tank (HEAT) and canister may be considered special types of bursting shell.
- Base Ejection: dual purpose improved conventional munitions (DPICM)-bomblet, scatterable mines, illuminating, coloured flare, smoke, incendiary, propaganda, chaff[18] (foil to jam radars: originally known as "window")[19] and modern exotics such as electronic payloads and sensor-fuzed munitions.
- Nose Ejection: shrapnel, flechette, star, incendiary.

Projectile stabilization

- **Rifled** Traditionally, artillery projectiles have been spin-stabilised, meaning that they spin in flight so that gyroscopic forces prevent them from tumbling. Spin is induced by gun barrels having rifling which engages a soft metal band around the projectile, called a "driving band" (UK) or "rotating band" (U.S.). The driving band is usually made of copper, but synthetic materials have also been used.
- **Smoothbore/Fin-Stabilized** In modern artillery smoothbore tubes have been used mostly by mortars. These projectiles use fins in the airflow at their rear to maintain correct orientation. The primary benefit over rifled barrels is reduced barrel wear and longer ranges that can be achieved (due to the reduced loss of energy to friction and gas escaping around the projectile via the rifling).
- **Rifled/Fin-Stabilized** A combination of the above can be used, where the barrel is rifled, but the projectile also has deployable fins for stabilization,[20] guidance[21] or gliding.[22]

Propellant

Most forms of artillery require a propellant to propel the projectile at the target. Propellant is always a low explosive, this means it deflagrates instead of detonating, as with high explosives. The shell is accelerated to a high velocity in a very short time by the rapid generation of gas from the burning propellant. This high pressure is achieved by burning the propellant in a contained area, either the chamber of a gun barrel or the combustion chamber of a rocket motor.

152 mm howitzer D-20 during the Iran-Iraq war.

Until the late 19th century the only available propellant was black powder. Black powder had many disadvantages as a propellant; it has relatively low power, requiring large amounts of powder to fire projectiles, and created thick clouds of white smoke that would obscure the targets, betray the positions of guns and make aiming impossible. In 1846 nitrocellulose (also known as guncotton) was discovered, and the high explosive nitroglycerin was discovered at much the same time. Nitrocellulose was significantly more powerful than black powder, and was smokeless. Early guncotton was unstable however, and burned very fast and hot, leading to greatly increased barrel wear. Widespread introduction of smokeless powder would wait until the advent of the double-base powders, which combine nitrocellulose and nitroglycerin to produce powerful, smokeless, stable propellant.

Many other formulations were developed in the following decades, generally trying to find the optimum characteristics of a good artillery propellant; low temperature, high energy, non corrosive, highly stable, cheap, and easy to manufacture in large quantities. Broadly, modern gun propellants are divided into three classes: single-base propellants which are mainly or entirely nitrocellulose based, double-base propellants composed of a combination of nitrocellulose and nitroglycerin, and triple base composed of a combination of nitrocellulose and nitroglycerin and Nitroguanidine.

Artillery shells fired from a barrel can be assisted to greater range in three ways:

- rocket assisted projectiles (RAP) enhance and sustain the projectile's velocity by providing additional 'push' from a small rocket motor that is part of the projectile's base.
- Base bleed uses a small pyrotechnic charge at the base of the projectile to introduce sufficient combustion products into the low-pressure region behind the base of the projectile responsible for a large proportion of the drag.

- ramjet assisted, similar to rocket assisted but using a ramjet instead of a rocket motor; it is anticipated that a ramjet-assisted 120-mm mortar shell could reach a range of 22 mi (35 km).[23]

Propelling charges for tube artillery can be provided in one of two ways: either as cartridge bags or in metal cartridge cases. Generally anti-aircraft artillery and smaller caliber (up to 6" or 152.4 mm) guns use metal cartridge cases that include the round and propellant, similar to a modern rifle cartridge. This simplifies loading and is necessary for very high rates of fire. Bagged propellant allows the amount of powder to be raised or lowered depending on the range to the target. it also makes handling of larger shells easier. Each requires a totally different type of breech to the other. A metal case holds an integral primer to initiate the propellant and provides the gas seal to prevent the gases leaking out of the breech, this is called obturation. With bagged charges the breech itself provides obturation and holds the primer. In either case the primer is usually percussion but electrical is also used and laser ignition is emerging. Modern 155 mm guns have a primer magazine fitted to their breech.

Artillery ammunition has four classifications according to use:

Battleship Ammunition: 16" artillery shells aboard one of the United States *Iowa*-class battleships.

- **Service**: ammunition used in live fire training or for wartime use in a combat zone. Also known as "warshot" ammunition.
- **Practice**: Ammunition with a non- or minimally-explosive projectile that mimics the characteristics (range, accuracy) of live rounds for use under training conditions. Practice artillery ammunition often utilizes a colored-smoke-generating bursting charge for marking purposes in place of the normal high explosive charge.
- **Dummy**: Ammunition with an inert warhead, inert primer, and no propellant; used for training or display.
- **Blank**: Ammunition with live primer, greatly reduced propellant charge (typically black powder) and no projectile; used for training, demonstration or ceremonial use.

Field artillery system

Because field artillery mostly uses indirect fire the guns have to be part of a system that enables them to attack targets invisible to them in accordance with the combined arms plan.

Cyclone of the 320th French Artillery, in Hoogstade, Belgium, 5 September 1917.

The main functions in the field artillery system are:

- Communications
- Command: authority to allocate resources;
- Target acquisition: detect, identify and deduce the location of targets;
- Control: authority to decide which targets to attack and allot fire units to the attack;
- Production of firing data – to deliver fire from a fire unit onto its target;
- Fire units: guns, launchers or mortars grouped together;
- Specialist services – produce data to support the production of accurate firing data;
- Logistic services – to provide combat supplies, particularly ammunition, and equipment support.

Organisationally and spatially these functions can be arranged in many ways. Since the creation of modern indirect fire different armies have done it differently at different times and in different places. Technology is often a factor but so are military-social issues, the relationships between artillery and other arms, and the criteria by which military capability, efficiency and effectiveness are judged. Cost is also an issue because artillery is expensive due to the large quantities of ammunition that it uses and its level of manpower.

Communications underpin the artillery system, they have to be reliable and in real-time to link the various elements. During the 20th century communications used flags, morse code by radio, line and lights, voice and teleprinter by line. Radio has included HF, VHF, satellite and radio relay as well as modern tactical trunk systems. In western armies at least radio communications are now usually encrypted.

The emergence of mobile and man-portable radios after World War I had a major impact on artillery because it enable fast and mobile operations with observers accompanying the infantry or armoured troops. In World War II some armies fitted their self-propelled guns with radios. However, sometimes in the first half of the 20th century hardcopy artillery fire plans and map traces were distributed.

Data communications can be especially important for artillery because by using structured messages and defined data types fire control messages can be automatically routed and processed by computers. For example a target acquisition element can send a message with target details which is automatically routed through the tactical and technical fire control elements to deliver firing data to the gun's laying system and the gun automatically laid. As tactical data networks become pervasive they will provide any connected soldier with a means for reporting target information and requesting artillery fire.

Command is the authority to allocate resources, typically by assigning artillery formations or units. Terminology and its implications vary widely. However, very broadly, artillery units are assigned in direct support or in general support. Typically, the former mostly provide close support to manoeuvre units while the latter may provide close support and or depth fire, notably counter-battery. Generally, 'direct support' also means that the artillery unit provides artillery observation and liaison teams to the supported units. Sometimes direct support units are placed under command of the regiment/brigade they support. General support units may be grouped into artillery formations for example, brigades even divisions, or multi-battalion regiments, and usually under command of division, corps or higher HQs. General support units tend to be moved to where they are most required at any particular time. Artillery command may impose priorities and constraints to support their combined arms commander's plans.

Target acquisition can take many forms, it is usually observation in real time but may be the product of analysis. Artillery observation teams are the most common means of target acquisition. However, air observers have been use since the beginning of indirect fire and were quickly joined by air photography. Target acquisition may also be by anyone that can get the information into the artillery system. Targets may be visible to forward troops or in depth and invisible to them.

Observation equipment can vary widely in its complexity.

- Unmanned air vehicles are the latest form of air observation, having been first introduced in the early 1960s.
- The equipment available to observation teams has progressed from just prismatic compass, hand-held or tripod mounted binoculars and sometimes optical range-finders,
- Special equipment for locating hostile artillery: flash spotting and notably sound ranging appeared in World War I the latter has been undergone increasing refinement as technology has improved. These were joined by radar in World War II.
- In the mid-1970s several armies started equipping their artillery observation teams with laser rangefinders, ground surveillance radars and night vision devices, these were soon followed by inertial orienting and navigating devices to improve the accuracy of target locations. The Global Positioning System (GPS) provided a smaller and cheaper means of quick and accurate fixation for target acquisition devices.
- Specialised units with ground surveillance radars, unattended ground sensors or observation patrols operating in depth have also been used.
- Targets in depth may also be 'acquired' by intelligence processes using various sources and agencies such as HUMINT, SIGINT, ELINT and IMINT.
- Laser guided shells require laser target designators, usually with observation teams on the ground but UAV installations are possible.

- Specialised artillery observation vehicles appeared in World War II and have greatly increased in sophistication since that time.

Control, sometimes called tactical fire control, is primarily concerned with 'targeting' and the allotment of fire units to targets. This is vital when a target is within range of many fire units and the number of fire units needed depends on the nature of the target, and the circumstances and purpose of its engagement. Targeting is concerned with selecting the right weapons in the right quantities to achieve the required effects on the target. Allotment attempts to address the artillery dilemma—important targets are rarely urgent and urgent targets are rarely important. Of course importance is a matter of perspective; what is important to a divisional commander is rarely the same as what is important to an infantry platoon commander.

Broadly, there are two situations: fire against opportunity targets and targets whose engagement is planned as part of a particular operation. In the latter situation command assigns fire units to the operation and an overall artillery fire planner makes a plan, possibly delegating resources for some parts of it to other planners. Fire plans may also involve use of non-artillery assets such as mortars and aircraft.

Afghans with two captured artillery field guns in Jaji, 1984.

Control of fire against opportunity targets is an important differentiator between different types of artillery system. In some armies only designated artillery HQs have the tactical fire control authority to order fire units to engage a target, all 'calls for fire' being requests to these HQs. This authority may also extend to deciding the type and quantity of ammunition to be used. In other armies an 'authorised observer' (for example, artillery observation team or other target acquisition element) can order fire units to engage. In the latter case a battery observation team can order fire to their own battery and may be authorised to order fire to their own battalion and sometimes to many battalions. For example a divisional artillery commander may authorise selected observers to order fire to the entire divisional artillery. When observers or cells are not authorised they can still request fire.

Armies that apply forward tactical control generally put the majority of the more senior officers of artillery units forward in command observation posts or with the supported arm. Those that do not use this approach tend to put these officers close to the guns. In either case the observation element usually controls fire in detail against the target, such as adjusting it onto the target, moving it and co-ordinating it with the supported arm as necessary to achieve the required effects.

Firing data has to be calculated and is the key to indirect fire, the arrangements for this have varied widely. In the end firing data has two components: quadrant elevation and azimuth, to these may be added the size of propelling charge and the fuze setting. The process to produce firing data this is sometimes called technical fire control. Before computers, some armies set the range on the gun's sights, which mechanically corrected it for the gun's muzzle velocity. For the first few decades of indirect fire, the firing data were often calculated by the observer who then adjusted the fall of shot onto the target.

However, the need to engage targets at night, in depth or hit the target with the first rounds quickly led to predicted fire being developed in World War I. Predicted fire existed alongside the older method. After World War II predicted methods were invariably applied but the fall of shot usually needed adjustment because of inaccuracy in locating the target, the proximity of friendly troops or the need to engage a moving target. Target location errors were significantly reduced once laser rangefinders, orientation and navigation devices were issued to observation parties.

In predicted fire the basic geospatial data of range, angle of sight and azimuth between a fire unit and its target was produced and corrected for variations from the 'standard conditions'. These variations included barrel wear, propellant temperature, different projectiles weights that all affected the muzzle velocity, and air temperature, density, wind speed & direction and rotation of the earth that affect the shell in flight. The net effect of variations can also be determined by shooting at an accurately known point, a process called 'registration'.

All these calculations to produce a quadrant elevation (or range) and azimuth were done manually by highly trained soldiers using instruments, tabulated data, data of the moment and approximations until battlefield computers started appearing in the 1960s and 1970s. While some early calculators copied the manual method (typically substituting polynomials for tabulated data), computers use a different approach. They simulate a shell's trajectory by 'flying' it in short steps and applying data about the conditions affecting the trajectory at each step. This simulation is repeated until it produces a quadrant elevation and azimuth that lands the shell within the required 'closing' distance of the target co-ordinates. NATO has a standard ballistic model for computer calculations and has expanded the scope of this into the NATO Armaments Ballistic Kernel (NABK)[24] within the SG2 Shareable (Fire Control) Software Suite (S4).

Technical fire control has been performed in various places, but mostly in firing batteries. However, in the 1930s the French moved it to battalion level and combined it with some tactical fire control. This was copied by the US. Nevertheless most armies seemed to have retained it within firing batteries and some duplicated the technical fire control teams in a battery to give operational resilience and tactical flexibility. Computers reduced the number of men needed and enabled decentralisation of technical fire control to autonomous sub-battery fire units such as platoons, troops or sections, although some armies had sometimes done this with their manual methods. Computation on the gun or launcher, integrated with their laying system, is also possible. MLRS led the way in this.

A **fire unit** is the smallest artillery or mortar element, consisting of one or more weapon systems, capable of being employed to execute a fire assigned by a tactical fire controller. Generally it is a battery, but sub-divided batteries are quite common, and in some armies very common. On occasions a battery of 6 guns has been 6 fire units. Fire units may or may not occupy separate positions. Geographically dispersed fire units may or may not have an integral capability for technical fire control.

Members of the 3rd Marine Division complete construction of M101 105 mm howitzer positions at a mountain-top fire support base, Vietnam, 1968

Specialist services provide data need for predicted fire. Increasingly, they are provided from within firing units. These services include:

- Survey: accurate fixation and orientation of the guns, historically this involved specialists within field artillery units and specialist units. In some armies mapping and map supply has also been an artillery responsibility. Survey is also essential for some target acquisition devices. Traditional survey methods of measurement and calculation have been replaced by inertial orientation and navigators and GPS.
- Meteorological data: historically these were usually divisional level specialist teams but advances in technology mean they are now increasingly part of artillery units.
- Calibration: periodically establishing the "normal" muzzle velocity of each gun as it wears. Originally this involved special facilities and army level teams. Measurement using Doppler radar, introduced in the 1950s, started to simplify arrangements. Some armies now have a muzzle velocity measuring radar permanently fitted to every gun.

Logistic services, supply of artillery ammunition has always been a major component of military logistics. Up until World War I some armies made artillery responsible for all forward ammunition supply because the load of small arms ammunition was trivial compared to artillery. Different armies use different approaches to ammunition supply, which can vary with the nature of operations. Differences include where the logistic service transfers artillery ammunition to artillery, the amount of ammunition carried in units and extent to which stocks are held at unit or battery level. A key difference is whether supply is 'push' or 'pull'. In the former the 'pipeline' keeps pushing ammunition into formations or units at a defined rate. In the latter units fire as tactically necessary and replenish to maintain or reach their authorised holding (which can vary), so the logistic system has to be able to cope with surge and slack.

Artillery has always been equipment intensive and for centuries artillery provided its own artificers to maintain and repair their equipment. Most armies now place these services in specialist branches with specialist repair elements in batteries and units.

Classification of artillery

Artillery types can be categorised in several ways, for example by type or size of weapon or ordnance, by role or by organizational arrangements.

Types of ordnance

The types of cannon artillery are generally distinguished by the velocity at which they fire projectiles. Types of artillery:

German Army PzH 2000 self-propelled artillery

- Field artillery: mobile weapons used to support armies in the field. Subcategories include:
  - infantry support guns: directly support infantry units.
  - mountain guns: lightweight weapons that can be moved through difficult terrain.
  - field guns: capable of long range fire.
  - howitzers: capable of high angle fire, they are most often employed for indirect-fire.
  - gun howitzers: capable of high or low angle fire with a long barrel.
  - mortars: typically short-barreled, high-trajectory weapons designed primarily for an indirect-fire role.
  - anti-tank artillery: weapons, usually mobile, designed for attacking tanks.
  - anti-aircraft artillery: weapons, usually mobile, designed for attacking aircraft from the ground. Some guns were suitable for dual-role anti-aircraft and field (anti-tank) use. The World War II German 88 mm gun was a famous example.
  - rocket artillery: rocket-launched instead of shot or shell.
- Motorized artillery: towed by Artillery tractors or APU-installed.
- Self-propelled artillery: typically guns, mortars or gun howitzers mounted on a vehicle.
- Railway gun: large-caliber weapons that are mounted on, transported by and fired from specially-designed railway wagons.
- Naval artillery: guns mounted on warships and used either against other ships or in support of ground forces. The crowning achievement of naval artillery was the battleship, but the advent of airpower and missiles have rendered this type of artillery largely obsolete. They are typically longer-barreled, low-trajectory, high-velocity weapons designed primarily for a direct-fire role.
- Coastal artillery: Fixed-position weapons dedicated to defense of a particular location, usually a coast (for example, the Atlantic Wall in World War II) or harbor. Not needing to be mobile, coastal artillery used to be much larger than equivalent field artillery pieces, giving them longer range and more destructive power. Modern coastal artillery (for example,, Russia's "Bereg" system) is often self propelled, (allowing it to avoid counter-battery fire) and fully integrated, meaning that each battery has all of the support systems that it requires (maintenance, targeting radar, etc.) organic to its unit.

Modern field artillery can also be split into two other categories: towed and self-propelled. As the name suggests, towed artillery has a prime mover, usually a jeep or truck, to move the piece, crew, and ammunition around. Self-propelled howitzers are permanently mounted on a carriage or vehicle with room for the crew and ammunition and are thus capable of moving quickly from one firing position to another, both to support the fluid nature of modern combat and to avoid counter-battery fire. There are also mortar carrier vehicles, many of which allow the mortar to be removed from the vehicle and be used dismounted, potentially in terrain in which the vehicle cannot navigate, or in order to avoid detection.

Naval piece of artillery, early 19th century

Organizational types

At the beginning of the modern artillery period, the late 19th century, many armies had three main types of artillery, in some case they were sub-branches within the artillery branch in others they were separate branches or corps. There were also other types excluding the armament fitted to warships:

Australian gunners, wearing gas masks, operate a 9.2-inch (230 mm) howitzer during World War I.

- Horse artillery, first formed as regular units in late 18th century, with the role of supporting cavalry, they were distinguished by the entire crew being mounted.
- Field or "foot" artillery, the main artillery arm of the field army, using either guns, howitzers or mortars. In World War II this branch again started using rockets and later surface to surface missiles.
- Fortress or garrison artillery, manned a nation's fixed defences using guns, howitzers or mortars, either on land or coastal frontiers. Some had deployable elements to provide heavy artillery to the field army. In some nations coast defence artillery was a naval responsibility.
- Mountain artillery, a few nations treated mountain artillery as a separate branch, in others it was a speciality in another artillery branch. They used light guns or howitzers, usually designed for pack animal transport and easily broken down into small easily handled loads
- Naval artillery, some nations carried pack artillery on some warships, these were used and manhandled by naval (or marine) landing parties. At times, part of a ship's armament would be unshipped and mated to makeshift carriages and limbers for actions ashore, for example during the Second Boer War, during the First World War the guns from the stricken SMS *Königsberg* formed the main artillery strength of the German forces in East Africa.

After World War I many nations merged these different artillery branches, in some cases keeping some as sub-branches. Naval artillery disappeared apart from that belonging to marines. However, two new branches of artillery emerged during that war and its aftermath, both used specialised guns (and a few rockets) and used direct not indirect fire, in the 1950s and 1960s both started to make extensive use of missiles:

Firing of an 18-pound gun, Louis-Philippe Crepin, (1772–1851)

- Anti-tank artillery, also under various organisational arrangements but typically either field artillery or a specialist branch and additional elements integral to infantry, etc., units. However, in most armies field and anti-aircraft artillery also had at least a secondary anti-tank role. After World War II anti-tank in Western armies became mostly the responsibility of infantry and armoured branches and ceased to be an artillery matter, with some exceptions.

- Anti-aircraft artillery, under various organisational arrangements including being part of artillery, a separate corps, even a separate service or being split between army for the field and airforce for home defence. In some cases infantry and the new armoured corps also operated their own integral light anti-aircraft artillery. Home defence anti-aircraft artillery often used fixed as well as mobile mountings. Some anti-aircraft guns could also be used as field or anti-tank artillery, providing they had suitable sights.

However, the general switch by artillery to indirect fire before and during World War I led to a reaction in some armies. The result was accompanying or infantry guns. These were usually small, short range guns, that could be easily man-handled and used mostly for direct fire but some could use indirect fire. Some were operated by the artillery branch but under command of the supported unit. In World War II they were joined by self-propelled assault guns, although other armies adopted infantry or close support tanks in armoured branch units for the same purpose, subsequently tanks generally took on the accompanying role.

Equipment types

The three main types of artillery 'gun' are guns, howitzers and mortars. During the 20th century, guns and howitzers have steadily merged in artillery use, making a distinction between the terms somewhat meaningless. By the end of the 20th century, true guns with calibers larger than about 60 mm had become very rare in artillery use, the main users being tanks, ships, and a few residual anti-aircraft and coastal guns. The term "cannon" is a United States generic term that includes guns, howitzers and mortars, it is not used in other English speaking armies.

Continental Artillery crew from the American Revolution handling a cannon.

The traditional definitions differentiated between guns and howitzers in terms of maximum elevation (well less than 45° as opposed to close to or greater than 45°), number of charges (one or more than one charge), and having higher or lower muzzle velocity, sometimes indicated by barrel length. These three criteria give eight possible combinations, of which guns and howitzers are but two. However, modern 'howitzers' have higher velocities and longer barrels than the equivalent 'guns' of the first half of the 20th century.

True guns are characterised by long range, having a maximum elevation significantly less than 45°, a high muzzle velocity and hence a relatively long barrel, smooth bore (no rifling) and a single charge. The latter often led to fixed ammunition where the projectile is locked to the cartridge case. There is no generally accepted minimum muzzle velocity or barrel length associated with a gun.

Howitzers can fire at maximum elevations at least close to 45°, and up to about 70° is normal for modern ones. They also have a choice of charges, meaning that the same elevation angle of fire will achieve a different range depending on the charge used. They have rifled bores, lower muzzle velocities and shorter barrels than equivalent guns. All this means they can deliver fire with a steep angle of descent. Because of their multi-charge capability, their ammunition is mostly separate loading (the projectile and propellant are loaded separately).

A British 60-pounder (5-inch (130 mm)) gun at full recoil, in action during the Battle of Gallipoli, 1915. Photo by Ernest Brooks.

That leaves six combinations of the three criteria, some of which have been termed gun howitzers. A term first used in the 1930s when howitzers with a relatively high maximum muzzle velocities were introduced, it never became widely accepted, most armies electing to widen the definition of 'gun' or 'howitzer'. By the 1960s, most equipments had maximum elevations up to about 70°, were multi-charge, had quite high maximum muzzle velocities and relatively long barrels.

Mortars are simple, the modern mortar originated in World War I and there were several patterns. After that war, most mortars settled on the Stokes pattern, characterised by a short barrel, smooth bore, low muzzle velocity, generally firing at an elevation angle greater than 45°, and a very simple and light mounting using a 'baseplate' on the ground. The projectile with its integral propelling charge was dropped down the barrel from the muzzle to hit a fixed firing pin. Since that time, a few mortars have become rifled and adopted breech loading.

There are other recognised typifying characteristics for artillery. First the type of obturation used to seal the chamber and prevent gases escaping through the breech. This may use a metal cartridge case that also holds the propelling charge, a configuration called 'QF' or 'quickfiring' by some nations. The alternative does not use a metal cartridge case, the propellant being merely bagged or in combustible cases with the breech itself providing all the sealing. This is called 'BL" or 'breech loading' by some nations.

A second characteristic is the form of propulsion. Basically modern equipment can either be towed or self-propelled (SP). A towed gun fires from the ground and any inherent protection is limited to a gun shield. Towing by horse teams lasted throughout World War II in some armies, but others were fully mechanised with wheeled or tracked gun towing vehicles by the outbreak of that war. The size of a towing vehicle depends on the weight of the equipment and the amount of ammunition it has to carry.

A variation of towed is portee where the vehicle carries the gun which is dismounted for firing. Mortars are often carried this way. A mortar is sometimes carried in an armoured vehicle and can either fire from it or be dismounted to fire from the ground. Since the early 1960s it has been possible to carry lighter towed guns and most mortars by helicopter. Even before that, they were parachuted or landed by glider from the time of the first airborne trials in the USSR in the 1930s.

In an SP equipment, the gun is an integral part of the vehicle that carries it. SPs first appeared during World War I, but did not really develop until World War II. They are mostly tracked vehicles, but wheeled SPs started to appear in the 1970s. Some SPs have no armour and carry little or no ammunition. Armoured SPs usually carry a useful ammunition load. Early armoured SPs were mostly a 'casemate' configuration, in essence an open top armoured box offering only limited traverse. However, most modern armoured SPs have a full enclosed armoured turret, usually giving full traverse for the gun. Many SPs cannot fire without deploying stabilisers or spades, sometimes hydraulic. A few SPs are designed so that the recoil forces of the gun are transferred directly onto the ground through a baseplate. A few towed guns have been given limited self-propulsion by means of an auxiliary engine.

Two other forms of tactical propulsion were used in the first half of the 20th century: Railways or transporting the equipment by road, as two or three separate loads, with disassembly and re-assembly at the beginning and end of the journey. Railway artillery took two forms, railway mountings for heavy and super-heavy guns and howitzers and armoured trains as 'fighting vehicles' armed with light artillery in a direct fire role. Disassembled transport was also used with heavy and super heavy weapons and lasted into the 1950s.

Caliber categories

A third form of artillery typing is to classify it as 'light', 'medium', 'heavy' and various other terms. It appears to have been introduced in World War I, which spawned a very wide array of artillery in all sorts of sizes so a simple categorical system was needed. Some armies defined these categories by bands of calibers. Different bands were used for different types of weapons—field guns, mortars, anti-aircraft guns and coast guns.

Modern operations

List of countries in order of amount of artillery:[25]

1. Russia - 26,121[26]
2. Democratic People's Republic of Korea - 17,900+[27]
3. China - 17,700+[28]
4. India - 11,258+[29]
5. Republic of Korea - 10,774+[30]
6. United States - 8,137[31]
7. Turkey - 7,450+[32]
8. Israel - 5,432[33]
9. Egypt - 4,480[34]
10. Pakistan - 4,291+[35]

Two French Army Giat GCT 155mm (155mm AUF1) Self-propelled Guns, 40th Regiment d' Artillerie, with IFOR markings are parked at Hekon base, near Mostar, Bosnia-Herzegovina, in support of Operation Joint Endeavor.

Artillery is used in a variety of roles depending on its type and caliber. The general role of artillery is to provide *fire support*—"the application of fire, coordinated with the manoeuvre of forces to destroy, *neutralize* or *suppress* the enemy". This NATO definition, of course, makes artillery a supporting arm although not all NATO armies agree with this logic. The *italicised* terms are NATO's.[36]

Unlike rockets, guns (or howitzers as some armies still call them) and mortars are suitable for delivering *close supporting fire*. However, they are all suitable for providing *deep supporting fire* although the limited range of many mortars tends to exclude them from the role. Their control arrangements and limited range also mean that mortars are most suited to *direct supporting fire*. Guns are used either for this or *general supporting fire* while rockets are mostly used for the latter. However, lighter rockets may be used for direct fire support. These rules of thumb apply to NATO armies.

Modern mortars, because of their lighter weight and simpler, more transportable design, are usually an integral part of infantry and, in some armies, armor units. This means they generally do not have to *concentrate* their fire so their shorter range is not a disadvantage. Some armies also consider infantry operated mortars to be more responsive than artillery, but this is a function of the control arrangements and not the case in all armies. However, mortars have always been used by artillery units and remain with them in many armies, including a few in NATO.

In NATO armies artillery is usually assigned a tactical mission that establishes its relationship and responsibilities to the formation or units it is assigned to. It seems that not all NATO nations use the terms and outside NATO others are probably used. The standard terms are: *direct support*, *general support*, *general support reinforcing* and *reinforcing*. These tactical missions are in the context of the command authority: *operational command*, *operational control*, *tactical command* or *tactical control*.

In NATO direct support generally means that the directly supporting artillery unit provides observers and liaison to the manoeuvre troops being supported, typically an artillery battalion or equivalent is assigned to a brigade and its batteries to the brigade's battalions. However, some armies achieve this by placing the assigned artillery units under command of the directly supported formation. Nevertheless, the batteries' fire can be *concentrated* onto a single target, as can the fire of units in range and with the other tactical missions.

Application of fire

There are several dimensions to this subject. The first is the notion that fire may be against an *opportunity* target or may be *prearranged.* If it is the latter it may be either *on-call* or *scheduled.* Prearranged targets may be part of a *fire plan.* Fire may be either *observed* or *unobserved,* if the former it may be *adjusted,* if the latter then it has to be *predicted.* Observation of adjusted fire may be directly by a forward observer or indirectly via some other *target acquisition* system.

A 155 mm artillery shell fired by a United States 11th Marine Regiment M-198 howitzer

NATO also recognises several different types of fire support for tactical purposes:

- *Counterbattery fire*: delivered for the purpose of destroying or *neutralizing* the enemy's fire support system.
- *Counterpreparation fire*: intensive prearranged fire delivered when the imminence of the enemy attack is discovered.
- *Covering fire*: used to protect troops when they are within range of enemy small arms.
- *Defensive fire*: delivered by supporting units to assist and protect a unit engaged in a defensive action.
- *Final Protective Fire*: an immediately available prearranged barrier of fire designed to impede enemy movement across defensive lines or areas.
- *Harassing fire*: a random number of shells are fired at random intervals, without any pattern to it that the enemy can predict. This process is designed to hinder enemy forces' movement, and, by the constantly imposed stress, threat of losses and inability of enemy forces to relax or sleep, lowers their morale.
- *Interdiction fire*: placed on an area or point to prevent the enemy from using the area or point.
- *Preparation fire*: delivered before an attack to weaken the enemy position.

These purposes have existed for most of the 20th century, although their definitions have evolved and will continue to do so, lack of *suppression* in *counterbattery* is an omission. Broadly they can be defined as either:

- *Deep supporting fire*: directed at objectives not in the immediate vicinity of own force, for neutralizing or destroying enemy reserves and weapons, and interfering with enemy command, supply, communications and observation; or
- *Close supporting fire*: placed on enemy troops, weapons or positions which, because of their proximity present the most immediate and serious threat to the supported unit.

Two other NATO terms also need definition:

- *Neutralization fire*: delivered to render a target temporarily ineffective or unusable; and
- *Suppression fire*: that degrades the performance of a target below the level needed to fulfill its mission. Suppression is usually only effective for the duration of the fire.

USMC M-198 firing outside of Fallujah, Iraq in 2004

The tactical purposes also include various "mission verbs", a rapidly expanding subject with the modern concept of "effects based operations".

Targeting is the process of selecting target and matching the appropriate response to them taking account of operational requirements and capabilities. It requires consideration of the type of fire support required and the extent of coordination with the supported arm. It involves decisions about:

- what effects are required, for example, *neutralization* or *suppression*;

- the proximity of and risks to own troops or non-combatants;
- what types of munitions, including their fuzing, are to be used and in what quantities;
- when the targets should be attacked and possibly for how long;
- what methods should be used, for example, *converged* or *distributed*, whether adjustment is permissible or surprise essential, the need for special procedures such as precision or danger close
- how many fire units are needed and which ones they should be from those that are available (in range, with the required munitions type and quantity, not allotted to another target, have the most suitable line of fire if there is a risk to own troops or non-combatants);

The *targeting* process is the key aspect of tactical fire control. Depending on the circumstances and national procedures it may all be undertaken in one place or may be distributed. In armies practicing control from the front, most of the process may be undertaken by a forward observer or other target acquirer. This is particularly the case for a smaller target requiring only a few fire units. The extent to which the process is formal or informal and makes use of computer based systems, documented norms or experience and judgement also varies widely armies and other circumstances.

Surprise may be essential or irrelevant. It depends on what effects are required and whether or not the target is likely to move or quickly improve its protective posture. During World War II UK researchers concluded that for impact fuzed munitions the relative risk were as follows:

- men standing – 1
- men lying – 1/3
- men firing from trenches – 1/15–1/50
- men crouching in trenches – 1/25–1/100

Airburst munitions significantly increase the relative risk for lying men, etc. Historically most casualties occur in the first 10–15 seconds of fire, i.e. the time needed to react and improve protective posture, however, this is less relevant if airburst is used.

There are several ways of making best use of this brief window of maximum vulnerability:

- ordering the guns to fire together, either by executive order or by a "fire at" time. The disadvantage is that if the fire is *concentrated* from many dispersed *fire units* then there will be different times of flight and the first rounds will be spread in time. To some extent a large concentration offsets the problem because it may mean that only one round is required from each gun and most of these could arrive in the 15 second window.
- burst fire, a rate of fire to deliver three rounds from each gun within 10 or 15 seconds, this reduces the number of guns and hence fire units needed, which means they may be less dispersed and have less variation in their times of flight. Smaller caliber guns, such as 105 mm, have always been able to deliver three rounds in 15 seconds, larger calibers firing fixed rounds could also do it but it wasn't until the 1970s that a multi-charge 155 mm howitzer, FH-70 first gained the capability.
- multiple round simultaneous impact (MRSI).
- *time on target*, fire units fire at the time less their time of flight, this works well with prearranged scheduled fire but is less satisfactory for opportunity targets because it means delaying the delivery of fire by selecting a 'safe' time that all or most fire units can achieve. It can be used with both the previous two methods.

Counter-battery fire

Counter-battery fire is a type of mission assigned to military artillery forces, which are tasked with locating and firing upon enemy artillery. Typically, enemy artillery would be detected when they fire, and a counter-battery fire mission must proceed as quickly as possible before the enemy artillery battery finishes their salvo and relocates.

Originally, counter-battery fire relied on ground or air-based artillery observers noticing the source of the artillery fire (due to muzzle flashes, smoke, spotting the artillery pieces, etc.) and calculating firing solutions to strike back at them. Artillery spotting, along with reconnaissance, was one of the major roles for aircraft in warfare (see World War I). Modern counter-battery fire relies on counter-battery radar, which calculate the source of incoming artillery shells very accurately and quickly—so quickly, in fact, that return fire can sometimes begin before the first enemy shell or rocket has landed.

The development of fast and accurate counter-battery fire has led to the concept of shoot-and-scoot and concentration on the development of highly mobile artillery pieces (typically self-propelled guns like the US M109 Paladin, the South African G6 Howitzer or Soviet Army 2S1 Gvozdika, or rocket artillery like the Soviet Katyusha or the multi-national M270 MLRS). The idea is to fire and then move before any counter-battery fire can land on the original position.

The task of destroying enemy artillery batteries can also fall to attack aircraft, but unless they are already on patrol overhead, they are usually not quick enough to save friendly forces from damage. More often, ground-based counter-battery fire would suppress the enemy battery/batteries and force them to move, while aircraft would follow up later with a strike to destroy the rest of the enemy artillery.

Field artillery team

Modern artillery ammunition. Caliber 155mm as used by the PzH 2000

Modern field artillery (post–World War I) has three distinct parts: the forward observer (or FO), the fire direction center (FDC) and the actual guns themselves. The forward observer observes the target using tools such as binoculars, laser rangefinders, designators and call back fire missions on his radio, or relays the data through a portable computer via an encrypted digital radio connection protected from jamming by computerized frequency hopping.

The FO can communicate directly with the battery FDC, of which there is one per each battery of 4 8 guns. Otherwise the several FOs communicate with a higher FDC such as at a Battalion level, and the higher FDC prioritizes the targets and allocates fires to individual batteries as needed to engage the targets that are spotted by the FOs or to perform preplanned fires.

The Battery FDC computes firing data—ammunition to be used, powder charge, fuse settings, the direction to the target, and the quadrant elevation to be fired at to reach the target, what gun will fire any rounds needed for adjusting on the target, and the number of rounds to be fired on the target by each gun once the target has been accurately located—to the guns. Traditionally this data is relayed via radio or wire communications as a warning order to the guns, followed by orders specifying the type of ammunition and fuse setting, direction, and the elevation needed to reach the target, and the method of adjustment or orders for fire for effect (FFE). However in more advanced artillery units, this data is relayed through a digital radio link.

Other parts of the field artillery team include meteorological analysis to determine the temperature, humidity and pressure of the air and wind direction and speed at different altitudes. Also radar is used both for determining the location of enemy artillery and mortar batteries and to determine the precise actual strike points of rounds fired by battery and comparing that location with what was expected to compute a registration allowing future rounds to be fired with much greater accuracy.

Time on Target

A technique called Time on Target was developed by the U.S. Army during World War II. This technique uses a precise determination of the time of flight from each firing battery to the target area. When a Time on Target (TOT) is designated each battery that will join in firing on that target subtracts the time of flight from the TOT to determine the time to fire. Individual firing batteries train to fire their rounds as close to simultaneously as possible. When each firing battery fires their rounds at their individual time to fire every round will reach the target area nearly simultaneously. This is especially effective when combined with techniques that allow fires for effect to be made without preliminary adjusting fires.

A similar effect may be obtained by a single battery firing sequential rounds with different trajectories, with all rounds timed to arrive simultaneously.

MRSI

This is a modern version of the earlier "time on target" concept in which fire from different weapons was timed to arrive on target at the same time. It is possible for modern computer-controlled artillery to fire more than one volley at a target and have all the shells arrive simultaneously, which is called MRSI (Multiple Rounds Simultaneous Impact). This is because there is more than one trajectory for the rounds to fly to any given target: typically one is below 45 degrees from horizontal and the other is above it, and by varying the amount of propellant with each shell, it is possible to create multiple trajectories. Because the higher trajectories cause the shells to arc higher into the air, they take longer to reach the target and so if the shells are fired on these trajectories for the first volleys (starting with the shell with the most propellant and working down) and then after the correct pause more volleys are fired on the lower trajectories, the shells will all arrive at the same time. This is useful because many more shells can land on the target with no warning. With traditional volleys along the same trajectory, anybody at the target point will have a certain amount of time (however long it takes to reload and re-fire the guns) to run away or take cover between volleys. In addition, if guns in more than one location are firing on one target, with careful timing it can be arranged for all their shells to land at the same time for the same reason.

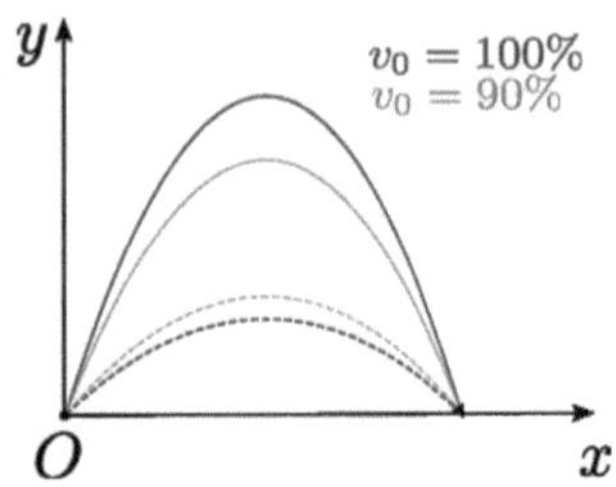

Illustration of different trajectories used in MRSI: For any muzzle velocity there is a steeper (> 45°, solid line) and a lower (<45°, dashed line) trajectory. On these different trajectories, the shells have different flight times.

Examples of MRSI guns are South Africa's Denel G6-52 (which can land six rounds simultaneously at targets at least 25 km (16 mi) away), Germany's Panzerhaubitze 2000 (which can land five rounds simultaneously at targets at least 17 km (11 mi) away) and Slovakia's 155 mm SpGH ZUZANA model 2000. The Archer project (Developed by BAE-Systems in Sweden), a 155 mm howitzer on a wheeled chassis claiming to be able to deliver up to 7 shells on target simultaneously from the same gun. The 120 mm twin barrel AMOS mortar system, developed in Finland, is capable of 7 + 7 shells MRSI. The United States Crusader program (now canceled) was slated to have MRSI capability.

Two-round MRSI firings were a popular artillery demonstration in the 1960s, where well trained crews could show off their skills for spectators.

Air burst

The destructiveness of artillery bombardments can be enhanced when some or all of the shells are set for airburst, meaning that they explode in the air above the target instead of upon impact. This can be accomplished either through time fuses or proximity fuses. Time fuses use a precise timer to detonate the shell after a preset delay. This technique is tricky and slight variations in the functioning of the fuse can cause it to explode too high and be ineffective, or to strike the ground instead of exploding above it. Since December 1944 (Battle of the Bulge), proximity fuzed artillery shells have been available that take the guesswork out of this process. These embody a miniature, low powered radar transmitter in the fuse to detect the ground and explode them at a predetermined height above it. The return of the weak radar signal completes an electrical circuit in the fuze which explodes the shell. The proximity fuse itself was developed by the British to increase the effectiveness of anti-aircraft warfare.

This is a very effective tactic against infantry and light vehicles, because it scatters the fragmentation of the shell over a larger area and prevents it from being blocked by terrain or entrenchments that do not include some form of robust overhead cover. Combined with TOT or MRSI tactics that give no warning of the incoming rounds, these rounds are especially devastating because many enemy soldiers are likely to be caught in the open. This is even more so if the attack is launched against an assembly area or troops moving in the open rather than a unit in an entrenched tactical position.

See also

- Artillery museums
- Gerald Bull, Project Babylon
- Gun laying
- List of artillery
- Nuclear artillery
- Paris Gun
- Henry Shrapnel – Inventor of "shrapnel shell"
- Suppressive fire
- Harry S. Truman a "redleg" in World War I

New artillery theories, concepts, methods and systems

- Advanced Gun System
- Combustion light gas gun
- Light Gas Gun
- Railgun

References

Notes

[1] Christopher Bellamy, Oxford Companion to Military History: artillery
[2] See siege engines for more information on pre-gunpowder devices.
[3] (http://www.guoxue123.com/shibu/0101/01ssqw/025.htm) History of Song
[4] Schmidtchen 1977, p. 162
[5] Cook, Weston F., Jr. *Warfare and Firearms in Fifteenth Century Morocco, 1400-1492.* 1993
[6] Clifford J. Rogers (1993) The Military Revolutions of the Hundred Years' War. The Journal of Military History, 57(2)April, 241–278 http://links.jstor.org/sici? sici=0899-3718(199304)57:2%3C241:TMROTH%3E2.0.CO;2-Fi, ISSN 08993718
[7] DeVries, K: The Use of Gunpowder Weaponry By and Against Joan or Arc During the Hundred Years' War. 1996
[8] Nicolle, David (2000). *Constantinople 1453: The end of Byzantium.* London: Osprey. pp. 29–30. ISBN 1-84176-091-9.
[9] Nicolle, David (1983). *Armies of the Ottoman Turks 1300-1774.* Osprey. pp. 29–30. ISBN 0-85045-511-1.
[10] Holmes, p.70
[11] Keegan, John, *A History of Warfare* (Vintage, 1993).
[12] Ordway, Vice-Commander of Artillery of the Polish king, Wladyslaw IV, *Great Art of Artillery, the First Part*, also known as *The Complete Art of Artillery*, pp.407–416.
[13] Christopher Bellamy, *Red God of War: Soviet Artillery and Rocket Forces*, London, 1986, p.16, quoted in Knox, MacGregor; Murray, Williamson (2001). *The Dynamics of Military Revolution.* New York: Cambridge University Press. p. 135. ISBN 052180079X.
[14] Knox, MacGregor; Murray, Williamson (2001). *The Dynamics of Military Revolution.* New York: Cambridge University Press. p. 136. ISBN 052180079X.
[15] Knox, MacGregor; Murray, Williamson (2001). *The Dynamics of Military Revolution.* New York: Cambridge University Press. p. 141. ISBN 052180079X.
[16] Jonathan B. A. Bailey (2004). *Field artillery and firepower.* Naval Institute Press. ISBN 1591140293
[17] 102.001 - PROXIMITY FUSE, Science Service Historical Image Collection (http://scienceservice.si.edu/pages/102001.htm)

[18] p.266, Browne & Thurbon
[19] p.262, International Aeronautic Federation
[20] "Fin-stabilized artillery shell" (http://www.patentstorm.us/patents/6779754-fulltext.html). patentstorm.us. Ausgust 24, 2004. .
[21] "Excalibur Precision Projectile" (http://www.globalsecurity.org/military/systems/munitions/m982-155.htm). Globalsecurity.org. .
[22] "Guided artillery missile with extremely long range" (http://www.patentstorm.us/patents/6748871-fulltext.html). patentstorm.us. Ausgust 24, 2004. .
[23] McNab, Chris; Hunter Keeter (2008). *Tools of Violence: Guns, Tanks and Dirty Bombs*. Osprey Publishing. p. 145. ISBN 1846032253.
[24] The public NABK Brochure NABK (https://aop37.lsec.dnd.ca/sites/s4_public_info/S4 Document Information/NABK brochure.pdf)
[25] Hackett, James, (ed.), *The Military Balance 2010*, The International Institute for Strategic Studies, 2010; Note: Only conventional tube ordnance is given, in use with land forces
[26] Hackett, James, (ed.), *The Military Balance 2010*, The International Institute for Strategic Studies, 2010, p.223; Note: the number given is for Land Forces only. Naval Infantry and Coastal Defense forces, Federal Border Guard Service, and Interior Troops use an additional 500+ ordinance pieces
[27] Hackett, James, (ed.), *The Military Balance 2010*, The International Institute for Strategic Studies, 2010, p.412
[28] Hackett, James, (ed.), *The Military Balance 2010*, The International Institute for Strategic Studies, 2010, p.400
[29] Hackett, James, (ed.), *The Military Balance 2010*, The International Institute for Strategic Studies, 2010, p.360
[30] Hackett, James, (ed.), *The Military Balance 2010*, The International Institute for Strategic Studies, 2010, p.414
[31] Hackett, James, (ed.), *The Military Balance 2010*, The International Institute for Strategic Studies, 2010, p.33; Note: the total is composed of 6,270+ ordnance used by the US Army, Army Reserve and National Guard with 1,867 used by the USMC
[32] Hackett, James, (ed.), *The Military Balance 2010*, The International Institute for Strategic Studies, 2010, p.165
[33] Hackett, James, (ed.), *The Military Balance 2010*, The International Institute for Strategic Studies, 2010, p.255
[34] Hackett, James, (ed.), *The Military Balance 2010*, The International Institute for Strategic Studies, 2010, p.248; Note: Syria, Egypt's strategic partner in the past wars against Israel, uses 3,440+ artillery pieces, and is the 11th ranking artillery user in the World
[35] Hackett, James, (ed.), *The Military Balance 2010*, The International Institute for Strategic Studies, 2010, p.368
[36] AAP-6(2006) NATO Glossary of Terms and Definitions.

Bibliography

- TM 9-2300
- Holmes, Richard (1988). *The World Atlas of Warfare: Military Innovations that Changed the Course of History*. New York: Viking Studio Books. ISBN 9780670819676. OCLC 17840438.
- McCamley, N J (2004). *Disasters Underground*. Barnsley: Pen & Sword Military. ISBN 9781844150229. OCLC 53241739.
- McNaughton, Andrew (January 1929). "The Development of Artillery in the Great War". *Canadian Defence Quarterly* **6** (2).
- Ordway, Frederick I (July 1970). "History of Astronautics Symposium: Mar Del Plata, Argentina, October 1969". *Technology and Culture* **11** (3). ISSN 0040-165X.
- Browne, J P R; Thurbon, M T (1998). *Electronic Warfare*. Brassey's air power, v. 4. London: Brassey's. ISBN 9781857531336. OCLC 38292289.
- International Aeronautic Federation (January–June 1977). *Interavia* (Geneva: Interavia SA) **32**: 262. ISSN 0020-5168.
- Schmidtchen, Volker (1977). "Riesengeschütze des 15. Jahrhunderts. Technische Höchstleistungen ihrer Zeit [Giant cannon of the 15th century: technical masterpieces of their era]" (in German). *Technikgeschichte* (Munich: Deutsches Museum) **44** (2): 153–173 (162–164). OCLC 85351643.

Further reading

- Hogg, Oliver Frederick Gillilan (1970). *Artillery: Its Origin, Heyday and Decline*. London: C. Hurst. ISBN 9780900966439. OCLC 99454.
- Bailey, J B A (2004). *Field Artillery and Firepower*. AUSA Institute of Land Warfare book. Annapolis, MD: Naval Institute Press. ISBN 9781591140290. OCLC 51931033.

External links

- Portsmouth Action Field Gun Pictures and Video (http://www.pafg.co.uk/)
- Naval Weapons of the World (http://www.navweaps.com/Weapons/index_weapons.htm)
- Cannon Artillery - The Voice of Freedom's Thunder (http://www.cannonartillery.com/)
- Modern Artillery (http://www.enemyforces.com/artillery.htm)
- Evans, Nigel F. (2001–2007) " British Artillery in World War 2 (http://nigelef.tripod.com/index.htm)"
- Artillery Tactics and Combat during the Napoleonic Wars (http://napoleonistyka.atspace.com/artillery_tactics.htm)
- Artillery of Napoleon's Imperial Guard (http://napoleonistyka.atspace.com/IMPERIAL_GUARD_artillery.htm)
- French artillery and its ammunition. 14th to the end of the 19th century (http://www.artillerie.info/)

Sri_Lanka_Army_Order_of_Precedence

Sri Lanka Army

Components

Sri Lanka Army

Sri Lanka Army Volunteer Force

List of regiments and corps

Structure of the Sri Lanka Army

History

History of the Sri Lanka Army

Equipment

Weaponry and equipment

Personnel

Commander of the Army

Army Ranks

For the purposes of parading, the regular army & the volunteer force of the Sri Lanka Army is listed according to an order of precedence.

- Sri Lanka Armoured Corps
- Sri Lanka Artillery
- Sri Lanka Engineers
- Sri Lanka Signals Corps
- Sri Lanka Light Infantry
- Sri Lanka Sinha Regiment
- Gemunu Watch
- Gajaba Regiment
- Vijayabahu Infantry Regiment
- Mechanized Infantry Regiment
- President's Guard
- Commando Regiment
- Special Forces Regiment
- Military Intelligence Corps
- Engineer Services Regiment
- Sri Lanka Army Service Corps
- Sri Lanka Army Medical Corps

- Sri Lanka Army Ordnance Corps
- Sri Lanka Electrical and Mechanical Engineers
- Sri Lanka Corps of Military Police
- Sri Lanka Army General Service Corps
- Sri Lanka Army Women's Corps
- Sri Lanka Rifle Corps
- Sri Lanka Army Pioneer Corps
- Sri Lanka National Guard

Sri_Lanka_Armed_Forces

Sri Lankan Armed Forces	

Service branches	 Sri Lanka Army  Sri Lanka Navy  Sri Lanka Air Force
Leadership	
Commander-in-Chief	President Mahinda Rajapaksa
Minister of Defence, Public Security, Law & Order	President Mahinda Rajapaksa
Manpower	
Military age	18 years of age
Conscription	None
Available for military service	5,342,147 males, age 16-49 (2010 est.), 5,466,409 females, age 16-49 (2010 est.)
Fit for military service	4,177,432 males, age 16-49 (2010 est.), 4,574,833 females, age 16-49 (2010 est.)
Reaching military age annually	167,026 males (2010 est.), 162,587 females (2010 est.)
Expenditures	
Budget	$1.28 billion (2010)
Percent of GDP	3.5% (2009)
Related articles	
History	Congo Crisis First JVP Insurrection Second JVP Insurrection Sri Lankan Civil War • Eelam War I • Eelam War II • Eelam War III • Eelam War IV

The **Sri Lanka Armed Forces** is the overall unified military of the Democratic Socialist Republic of Sri Lanka encompassing the Sri Lanka Army, the Sri Lanka Navy, the Sri Lanka Air Force which comes under preview of the

Ministry of Defence (MoD). The total strength of the three services numbers around 400,000 active personal who have voluntary joined, since military draft have never been imposed in Sri Lanka.

History

Sri Lanka has a long military history going back 2000 years. The roots of the modern Sri Lanka military leads back to the colonial era when Portuguese, Dutch and British established local militias to support their wars against the local Kingdoms. The British created the Ceylon Rifle Regiment during the Kandyan wars, although it had natives in its ranks majority of it was made up of Malays. It was disbanded 1873.

The linage of the Sri Lanka Armed Forces dates back to 1881, when the British creates a volunteer reserve in the island named the Ceylon Light Infantry Volunteers. Created to supplement the British garrison in Ceylon in the event of an external threat, it gradually increased in size. In 1910 it was renamed the Ceylon Defence Force (CDF) and consisted of several regiments. The CDF mobilized for home defence in World War I and again in World War II when its units were deployed along with allied forces in Asia and Africa. At the end of the war it has grown in size of an independent brigade, but was de-mobilized in 1946 and disbanded in 1949. In 1937 the *Ceylon Naval Volunteer Force* was established (later renamed as the *Ceylon Royal Naval Volunteer Reserve* (CRNVR)), it mobilized for war in 1939 and was taken into the Royal Navy.

Following establishment of the **Dominion of Ceylon** with Britain granting Independence in 1948, work began to establish a regular military. The Army Act No. 17 of 1949 which was passed by Parliament on April 11, 1949 and formalized in Gazette Extraordinary No. 10028 of October 10, 1949 marked the creation of the Ceylon Army, as the CDF the CRNVR disbanded to make way for a regular navy when the December 9, 1950 the Royal Ceylon Navy (RCyN) was established with its personal. Final in 1951 as the Royal Ceylon Air Force (RCyAF) was established as the youngest of the three forces. From the out set Britain played a significant role in helping the Ceylon government in developing its armed forces.

The growth for the Ceylon Armed Forces were slow due to lack of foreign threat, since Ceylon maintained cordial relations with its powerful neighbor India and the defence treaty with Britain. In the 1950s it was mainly employed in internal security assisting the police. There was an attempted coup in 1962 by a group of reservist, which led to cuts in military spending and the disbandment of several regiments. This together with the lack of an intelligence agency left it ill prepared for the insurgency launched by the Marxist JVP in April 1971.

The 1971 JVP Insurrection saw Ceylon Armed Forces mobilizing for combat operations for the first time and its size grew rapidly. The insurrection was brought under control in a few months. In 1972 Ceylon became a Republic and the Ceylon Armed Forces became the Sri Lanka Armed Forces.

By the early 1980s, the Sri Lanka Armed Forces mobilizing again as a new insurgency began by Tamil militant groups in the north of the island. This was the beginning of the Sri Lankan Civil War. The size of the Armed Forces grew at a rapid rate in the 1980s. By the mid-1980s, the Armed Forces began launching operations in the like of conventional warfare against the LTTE which had by then became the most powerful of the Tamil militant groups. This led to India intervening by breaching Sri Lankan air space to carry out food drops and shortly afterward the Indo-Sri Lankan Accord was signed and the Indian Peace Keeping Force (IPKF) was sent to Sri Lanka to establish peace.

The military was redistricted to it bases and was soon heavily involved in another insurrection by the JVP in the south of the island from 1987 to 1989. In the north tension increased with the LTTE and the IPKF leading to open war with the two suffering heavy casualties. In 1990 the IPKF pulled out and the war commenced with the Sri Lanka Armed Forces and the LTTE.

In 1994, a brief ceasefire came into place and peace talks began. But it broke down when the LTTE attacked several docked naval gunboats. The phase of the war that followed; known as Eelam War III saw a conventional war taking place in the northern and eastern provinces of the island and terrorist attacks in other parts of the country. The Sri

Lankan Army began deploying full divisions in offensive operations and Navy and Air Force increased it inventory to support the army.

In 2002, a new ceasefire was established with Norwegian mediation and Peace talks began. The SLMM was established to monitor the ceasefire and certain progress archived until the LTTE withdrew from the peace talks in 2003. Although the ceasefire continued no peace talks took place till 2005. In the mean time the Sri Lanka Monitoring Mission (SLMM) filed a report stating the LTTE had violated the ceasefire 3,471 since the signing of the cease fire which included matters like child recruiting, torture, abduction, firing of weapons, sabotage, carrying of arms in government-held areas, construction of new positions, movement of arms, ammunition and military equipment, families of detainees denied access and the obstruction of truce monitors.[1] However the security forces violated the ceasefire only 162 times.

In April 2006 following the a suicide bomb attack on the Commander of the Army, air strikes began followed by skirmishers, however both the government and the LTTE claimed that the ceasefire was still in place. Offensive by the Armed Forces was launched when the LTTE closed the sluice gates of the Mavil Aru reservoir on July 21 and cut the water supply to 15,000 villages in government controlled areas. This led to several major attacks by the LTTE in the eastern province and the north. The Armed Forces went on the offensive successfully liberating LTTE control areas in the eastern province during 2007.

On the 3 January 2008 the government informed Norway of its decision to quit the ceasefire, with it the ceasefire officially ended on the 16 January 2008, following several bombings in the capital. Along with the ceasefire ended the operations of the SLMM.[2]

During 2008 there were heavy fighting in the northern province where the Sri Lanka Armed Forces launched major offensives and succeeded in liberating LTTE controlled areas of the Mannar District, the Vavuniya District and moving in to the Mullaitivu District and Kilinochchi District. During December 2008, the Sri Lankan Armed Forces were engaged in offensives in all fronts, with heavy fighting around the Kilinochchi town where the LTTE had their headquarters and close to the Mullaitivu town.

In early 2009 the Armed Forces liberated in quick succession Kilinochchi town and the strategically important Elephant Pass. Thus establishing a land route to the government controlled Jaffna Peninsula which had been supplied by sea and air for over 10 years after its liberation in 1995. Shortly thereafter Mullaitivu town was liberated by the 59 Division of the SLA. Boxed into a small land area north of Mullaitivu, the LTTE with its remaining carders and leadership was effectively trapped, with this land mass been slowly reduced until May 2009.

On 19 May 2009, Sri Lanka Armed Forces won its final battle against the LTTE with the death of several top LTTE leaders, including its head Velupillai Prabhakaran while attempting to flee.[3]

On 22 May 2009, Sri Lankan Defence Secretary Gotabhaya Rajapaksa confirmed that 6,261 personnel of the Sri Lankan Armed Forces had lost their lives and 29,551 were wounded during Eelam War IV since July 2006.[4]

The Armed Forces have been accused of human rights violations during the war, particularly during the final stages. The Times alleged that United Nations reports indicate more than 20,000 civilians had died due to attacks carried out by the military.[5] However, the UN disputed these figures but admitted that the death toll was unacceptably high.[6] The Sri Lankan government strongly denied these figures.[7] The Human Rights Watch has also accused the military of "impeding the delivery of human aid" and carrying out indiscriminate attacks against civilians.[8]

Strategic Importance

The posture of the military has been defensive due to the nature of the strategic threats to Sri Lanka. In the short-term, internal security is considered the main threat to the nation's future. In the long-term, the threat is seen as primarily external from current and future superpowers in their rival quests for dominance of the Indian Ocean; at one point these were the United States of America and the Union of Soviet Socialist Republics. However, due to collapse of the Soviet Union in 1991, the external threats now involve future superpowers India and China. The signing of the Indo-Sri Lanka Peace Accord eased these concerns.[9]

The military of Sri Lanka has received a significant amount of military aid from the People's Republic of China.[10][11] Throughout the period from 2005 onwards, as a symbol of goodwill and cooperation between the two countries in terms of security, China has supplied the Sri Lankan army with Chinese military equipment.[12] Weapons provided by China include small arms, tanks,[13] and howitzers, as well as mobile vehicles. This can be seen as China's attempt to gain leverage against other power blocs, mainly India, within the Indian Ocean.[14] Sri Lanka also imports high-tech weapon systems from Israel.

Command organization

As head of state, the President of Sri Lanka, is nominally the Commander-in-Chief of the Armed Forces. The National Security Council, chaired by the President is the authority charged with formulating and executing defence policy for the nation. The highest level of military headquarters is the Ministry of Defence, since 1978 except for a few rare occasions the President retained the portfolio defence, thus being the Minister of Defence. The ministry and the armed forces have been controlled by the during these periods by either a Minister of State, Deputy Minister for defence, and of recently the Permanent Secretary to the Ministry of Defence. Prior to 1978 the Prime Minister held the portfolio of Minister of Defence and External Affairs, and was supported by a Parliamentary Secretary for Defence and External Affairs.

The Ministry of Defence is responsible for the management of the forces, while the planning and execution of combined operations is the responsibility of the Joint Operations Command (JOC). The JOC is headed by the Chief of the Defence Staff who is the most senior officer in the Armed Forces and is an appointment that can be held by an Air Chief Marshal, Admiral, or General. The three services have their own respective professional chiefs: the Commander of the Army, the Commander of the Navy and the Commander of the Air Force, who have much autonomy.

Armed Services

Army

The Sri Lanka Army is the oldest and largest of Sri Lanka's three armed services. Established as the Royal Ceylon Army in 1949, it was renamed when Sri Lanka became a republic in 1972. The Army of approximately 203,000 regular and reserve personnel including 73,000 National Guardsmen and is responsible for overseeing land-based military and humanitarian operations.

At present the army has deployed 12 divisions and forming 4 more in while carrying out combat operations. Since 2004 the Sri Lankan Army maintains a battalion and support units for 1000 personnel in the United Nations peacekeeping operations in Haiti.

An Armored Force of the Sri Lanka Army

Navy

The Sri Lankan Navy with approximately 48,000 personal is the key maritime division of the Sri Lankan Armed Forces and is classed as the most vital defence force of Sri Lanka. It conducts maritime operations at sea for the defence of the Sri Lankan nation and its interests. The professional head of the navy is the Commander of the navy, who exercises his command from the Naval Headquarters in Colombo. Established in 1950 as the Royal Ceylon Navy it was renamed as the Sri Lanka Navy in 1972.

In recent years it had played a key role it the Sri Lankan civil war, conducting deep sea, costal & inshore patrols, amphibious and supply operations. The navy has its own elite naval special forces unit, the Special Boat Squadron.

A IAI Kfir of the No. 10 "Fighter" Squadron

Air Force

The Sri Lanka Air Force with approximately 28,700 personal is the aerial defense division and the youngest of the Sri Lankan Tri Forces. Founded in 1951 as the Royal Ceylon Air Force, it relied on the British Royal Air Force for its earliest equipment, training, and leadership. The Air Force is playing a major role in the ongoing war against Tamil separatists. Although Sri Lanka is only a small island state, its Air Force has proven highly capable and efficient. In recent times the air force has expanded to specialize mainly in providing air-support to ground forces and landing troops and carrying airstrikes on rebel held areas in the Northern and Eastern provinces.

Paramilitary Forces

There are three official paramilitary forces under the command of the Ministry of Defence:

- The Special Task Force comes under the control of the Sri Lanka Police which in turn is under the command of the Ministry of Defence. It undertakes counter-terrorist operations and VIP protection.[15]
- The Sri Lanka Civil Defence Force, which is under direct command of the Ministry of Defence.[16]
- The Sri Lanka Coast Guard, which is under direct command of the Ministry of Defence.

See also Paramilitary groups of Sri Lanka

Training

Today the training of all armed services are carried out in Sri Lanka. With a Defence University, a Staff College, three Military Academies for the three armed services and many specialized training schools and centers members of the Sri Lanka armed forces have a high level training and professionalism.

- General Sir John Kotelawala Defence University, Colombo
- Defence Services Command and Staff College
- Sri Lanka Military Academy, Diyatalawa
- Naval and Maritime Academy, Trincomalee
- Air Force Academy, SLAF China Bay, Trincomalee

The Sri Lankan military has received specialized training assistance from other nations such as Pakistan, Australia, the United Kingdom, Israel, the United States, India, and South Africa.[9]

Referring to the overseas military training given to Sri Lanka, the U.S. government's contribution at a higher level. The *Foreign Military Financing (FMF)* and *International Military Education and Training (IMET)* programs helping to standardizing and strengthening the country's military greatly.[17]

Development

Although much of the current military hardware used by the Sri Lanka Armed Forces is acquired from China, Pakistan, Israel, India, Russia, and the United States indigenous weapon systems have been developed and produces within Sri Lanka to suit its requirements. Most of these have been produced by the armed forces. Following are military hardware design and development in the country:

- Unibuffel Mine-protected armoured personnel carrier[18]
- Unicorn Mine-protected armoured personnel carrier
- Jayasagara class Fast gun boat
- 29 M Landing Craft
- Colombo class Fast attack craft
- Arrow class combat speed boat

Awards & Decorations

The most prestigious decoration is Parama Weera Vibhushanaya which is awarded to officers, soldiers, sailors and airmen who have shown individual gallantry in the face of danger and risk their own lives to save the lives of their comrades. The armed forces especially the Army awards campaign medals for soldiers that have taken part in successful, very high-risk operations such as Operation Riviresa.

War crimes

There are allegations that war crimes were committed by the Sri Lankan military and the rebel Liberation Tigers of Tamil Eelam (Tamil Tigers) during the Sri Lankan Civil War, particularly during the final months of the conflict in 2009. The alleged war crimes include attacks on civilians and civilian buildings by both sides; executions of combatants and prisoners by both sides; enforced disappearances by the Sri Lankan military and paramilitary groups backed by them; acute shortages of food, medicine, and clean water for civilians trapped in the war zone; and child recruitment by the Tamil Tigers.[19] [20]

A United Nations expert panel investigating violations of international human rights and humanitarian law during the final stages of the civil war found "credible allegations" which, if proven, indicated that war crimes and crimes against humanity were committed by the Sri Lankan military and the Tamil Tigers.[21] [22] [23] The panel has called on the UN Secretary General to conduct an independent international inquiry into the alleged violations of international law.[24] [25] The Sri Lankan government has denied that its forces committed any war crimes and has strongly opposed any investigation.

Non-military activities

Since the end of the civil war in May 2009 Sri Lanka's 300,000 strong military has increased its non-military activities, leading to accusations of militarisation and even military rule.[26] [27] [28] The military is involved in everything from large scale property developments to the running of roadside cafes. The military has built roads, bridges, houses and stadiums.[26] [29] The Urban Development Authority, which spends hundreds billions of rupees a year, was taken over by the Ministry of Defence in 2010.[30] [31] A year later the ministry was renamed Ministry of Defence and Urban Development.[32] The ministry will have a budget of 229.9 billion rupees ($2.1 billion) in 2012, the largest of any government ministry.[33]

In early 2011, as food prices soared, the Army bought vegetables from farmers and then sold them to the public at a below market prices.[26] [29] [34] Some of the vegetables came from military farms.[26] The Army also runs a travel agency called *Air Travel Services (Pvt) Ltd* selling air-tickets and foreign package holidays.[26] [29] [31] In the war affected north of the country, where the military presence is significantly higher than the rest of the country even after the end of the conflict, the military operates numerous small restaurants and shops catering mainly for tourists from the south.[31] [35] Most of the the roadside cafes along the the A9 highway are owned and operated by the Army.[26] The Army runs a luxurious holiday resort called *Thalsevana* near Kankesanthurai inside the Valikamam North High Security Zone where the residents were forcibly expelled in the early 1990s and have not been allowed to resettle despite the end of the civil war.[26] [29] [31] [36] [37] The Army also runs two other holiday resorts, one in Kukuleganga and the other in Wadduwa.[26] [37] In November 2011 it was announced that the Army would build a five star hotel in Colombo.[38] During 2010-11, whilst local government was suspended in Colombo using emergency regulations, the Army took over many of the functions of the Colombo Municipal Council such as supervision of the collection of rubbish, demolition of low income housing and even decoration of the streets during festivals.[26] [29] [39]

The Navy's *Jetliner* (A542) passenger ship is used as a recreational cruise ship providing onboard weddings, receptions, corporate events, parties and other social functions.[40] The Navy's A543 vessel provides a whale and dolphin watching service for tourists.[26] [29] [31] [37] [41] The Navy also provides a canal-boat service in Colombo from Wellawatta to Nawala, a boat service to Adam's Bridge for tourists and a vegetable shop.[26] [29] [37] Helitours is the commercial arm of the Air Force established in the 1970s.[42] It was inactive during the civil war but after its end the business has started functioning again, taking advantage of the country's booming tourism.[27] [40]

From 2011 all undergraduates are required to take part in compulsory leadership training and the government has chosen the military to provide this training.[26] [29] [27] [43] [44] [45] In November 2011 it was announced that the miltary would take over the maintenance of the country's three international cricket stadiums from the financially troubled national cricket board.[27] [46] The Army has taken over the Hambantota Cricket Stadium, the Navy Pallekele Cricket Stadium and the Air Force Premadasa Stadium.[27] [47]

Many senior military commanders have been appointed to senior positions in the diplomatic and civil services: Air Chief Marshal Jayalath Weerakkody is the High Commissioner to Pakistan; Air Chief Marshal Donald Perera is the Ambassador to Israel; Major General Nanda Mallawaarachchi is the Ambassador to Indonesia; Major General Udaya Perera is the Deputy High Commissioner to Malaysia; Major General Jagath Dias is the Deputy Ambassador to Germany, Switzerland and the Vatican; Major General Shavendra Silva is the Deputy Permanent Representative to the UN; Admiral Wasantha Karannagoda is the Ambassador to Japan; Admiral Thisara Samarasinghe is the High Commissioner to Australia; Major General Amal Karunasekara is the charge d'affaires in Eritrea; Major General G. A. Chandrasiri is the Governor of Northern Province; Mohan Wijewickrema s the Governor of Eastern Province; Rohan Daluwatte is the Chairman of the National Gem and Jewellery Authority; and Lieutenant General Jagath Jayasuriya and Air Chief Marshal Roshan Goonetilleke are board members of the Water's Edge Complex.[48]

See also

- Sri Lanka
- Ministry of Defence (Sri Lanka)
- Sri Lankan Civil War

References

[1] Geneva talks must focus on SLMM rulings on violations (http://www.lankanewspapers.com/news/2006/2/5577.html)

[2] CFA, SLMM cease to operate by Jan 16, Defence.lk (http://www.defence.lk/new.asp?fname=20080104_02)

[3] The Last Day of Prabhakaran, D.B.S. Jeyaraj, Daily Mirror (http://www.dailymirror.lk/DM_BLOG/Sections/frmNewsDetailView.aspx?ARTID=49739,)

[4] Reuters article citing Gotabhaya Rajapaksa (http://www.reuters.com/article/latestCrisis/idUSSP463682,)

[5] Philp, Catherine (2009-05-29). "The hidden massacre: Sri Lanka's final offensive against Tamil Tigers" (http://www.timesonline.co.uk/tol/news/world/asia/article6383449.ece). *The Times*. . Retrieved 2009-07-24.

[6] "Ban denies UN cover-up of Sri Lanka death toll" (http://www.taipeitimes.com/News/world/archives/2009/06/03/2003445198). Taipei Times. 2009-06-03. . Retrieved 2009-07-24.

[7] "Sri Lanka rejects deaths report" (http://news.bbc.co.uk/2/hi/south_asia/8073540.stm). *BBC*. 2009-05-29. . Retrieved 2009-07-24.

[8] "Sri Lanka: Government Abuses Intensify" (http://www.hrw.org/en/news/2007/08/05/sri-lanka-government-abuses-intensify). Human Rights Watch. 2007-08-05. . Retrieved 2009-07-24.

[9] Sri Lanka, PRIMARY THREATS TO NATIONAL SECURITY (http://lcweb2.loc.gov/cgi-bin/query/r?frd/cstdy:@field(DOCID+lk0155))

[10] ""! (http://www.chnqiang.com/article/2009/0317/article_73297.shtml)

[11] (http://www.1n0.net/Article/ggjq/63417.html)

[12] (http://www.mitbbs.com/article/Military/31677108_0.html)

[13] (http://www.1n0.net/Article/ggjq/64327.html)

[14] , (http://www.1n0.net/Article/ggjq/64372.html)

[15] Task Force, Sri Lanka Police Service (http://www.police.lk/divisions/stf.aspSpecial)

[16] Home Guard Service, Ministry of Defence (http://www.defence.lk/main_abt.asp?fname=homeguard)

[17] "Sri Lanka: Security Assistance" (http://www.state.gov/t/pm/64481.htm). U.S. Department of State. July 12, 2007. .

[18] Nonis, Anton (2004-04-25). "The evolution of the indigenous armoured vehicle: From Unicorn to Unibuffel" (http://www.sundayobserver.lk/2004/04/25/fea22.html). The Sunday Observer. . Retrieved 2007-03-16.

[19] "Sri Lanka: US War Crimes Report Details Extensive Abuses" (http://www.hrw.org/en/news/2009/10/22/sri-lanka-us-war-crimes-report-details-extensive-abuses). Human Rights Watch. 22 October 2009. . Retrieved 17 January 2010.

[20] "Govt.: LTTE Executed Soldiers" (http://www.thesundayleader.lk/2010/12/08/govt-ltte-executed-soldiers/). The Sunday Leader. 8 December 2010. . Retrieved 17 January 2010.

[21] "Report of the UNSG's panel of experts on accountability in SL" (http://www.island.lk/index.php?page_cat=article-details&page=article-details&code_title=23146). *The Island, Sri Lanka*. 16 April 2011. .

[22] "UN panel admits international failure in Vanni war, calls for investigations" (http://www.tamilnet.com/art.html?catid=79&artid=33808). *TamilNet*. 16 April 2011. .

[23] "Summary of UN Panel report" (http://www.dailymirror.lk/top-story/10913-summary-of-un-panel-report.html). *Daily Mirror (Sri Lanka)*. 16 April 2011. .

[24] "Sri Lankan military committed war crimes: U.N. panel" (http://www.thehindu.com/news/international/article1701700.ece). *The Hindu*. 16 April 2011. .

[25] "Leaked UN report urges Sri Lanka war crimes probe" (http://www.france24.com/en/20110416-leaked-un-report-urges-sri-lanka-war-crimes-probe). *France24*. 16 April 2011. .

[26] "In bigger barracks" (http://www.economist.com/node/18775481). *The Economist*. 2 June 2011. .

[27] "Lanka hands over upkeep of stadiums to military" (http://www.gulf-times.com/site/topics/article.asp?cu_no=2&item_no=468279&version=1&template_id=44&parent_id=24). *Gulf Times*. 5 November 2011. .

[28] Gunasekara, Tisaranee (6 November 2011). "Rewarding Illegality; Enthroning Injustice" (http://www.thesundayleader.lk/2011/11/06/rewarding-illegality-enthroning-injustice/). *Sunday Leader*. .

[29] Bedi, Rahul (20 August 2011). "Armed, dangerous and building their own empires" (http://www.nzherald.co.nz/business/news/article.cfm?c_id=3&objectid=10746321). *The New Zealand Herald*. .

[30] Hemmathagama, Ashwin (23 October 2011). "Government May Borrow More!" (http://www.thesundayleader.lk/2011/10/23/government-may-borrow-more/). *Sunday Leader*. .

[31] Haviland, Charles (14 March 2011). "Sri Lanka army opens travel wing" (http://www.bbc.co.uk/sinhala/news/story/2011/03/110314_army_travel_agency.shtml). *BBC Sinhala*. .

[32] "Defence Ministry adds UD to its name" (http://www.ft.lk/2011/10/17/defence-ministry-adds-ud-to-its-name/). *Daily FT*. 17 October 2011. .

[33] "Estimated income - Rs. 1115 b., expenditure –Rs. 2220 b." (http://sundaytimes.lk/111016/News/nws_14.html). *Sunday Times (Sri Lanka)*. 16 October 2011. .

[34] Haviland, Charles (7 January 2011). "Sri Lanka army becomes vegetable vendor as prices rise" (http://www.bbc.co.uk/news/world-south-asia-12138540). *BBC News*. .

[35] Bateman, Gibson (22 October 2011). "The Continued Militarization of Sri Lanka" (http://www.jofr.org/2011/10/22/the-continued-militarization-of-sri-lanka/). *Journal of Foreign Relations*. .

[36] "SLA opens 'luxury' resort in occupied Tamil land" (http://www.tamilnet.com/art.html?catid=13&artid=32744). *TamilNet*. 5 October 2011. .

[37] Jayasuriya, Ranga. "Army's new civilian role - is this a good signal?" (http://lakbimanews.lk/index.php?option=com_content&view=article&id=1460:armys-new-civilian-role-is-this-a-good-signal&Itemid=56). *Lakbima*. .

[38] Dias, Supun (6 November 2011). "Army to build a five star hotel in Colombo" (http://www.dailymirror.lk/news/14575-army-to-build-a-five-star-hotel-in-colombo.html). *Daily Mirror (Sri Lanka)*. .

[39] "RESTORE CIVIL ADMINISTRATION IN ALL ASPECTS TO ENSURE DEMOCRACY" (http://www.peace-srilanka.org/~peace/index.php?option=com_content&view=article&id=385:restore-civil-administration-in-all-aspects-to-ensure-democracy&catid=17:media). The National Peace Council of Sri Lanka. 1 June 2011. .

[40] Samarajiva, Indi (4 April 2010). "Military Tourism" (http://www.thesundayleader.lk/2010/04/04/military-tourism/). *Sunday Leader*. .

[41] "Ripples as Navy sails into whale-watching business" (http://sundaytimes.lk/110130/News/nws_02.html). *Sunday Times (Sri Lanka)*. 30 January 2011. .

[42] "Helitours dressed up; ready to go as a commercial venture" (http://sundaytimes.lk/100718/BusinessTimes/bt02.html). *Sunday Times (Sri Lanka)*. 18 July 2010. .
[43] Abeywickrema, Mandana Ismail (24 April 2011). "University Students To Be Trained By Military" (http://www.thesundayleader.lk/2011/04/24/university-students-to-be-trained-by-military/). *Sunday Leader*. .
[44] Haviland, Charles (19 May 2011). "Sri Lanka students fight military training scheme" (http://www.bbc.co.uk/news/world-south-asia-13461858). *BBC News*. .
[45] Hensman, Rohini (14 June 2011). "Sri Lankan universities are no place for the army" (http://www.guardian.co.uk/commentisfree/2011/jun/14/sri-lanka-army-university-training). *The Guardian*. .
[46] Haviland, Charles (3 November 2011). "Sri Lanka army to run cricket stadiums" (http://www.bbc.co.uk/news/world-asia-15574128). *BBC News*. .
[47] "Three Test venues handed to military for maintenance, confirms Sports Ministry" (http://www.sundaytimes.lk/index.php?option=com_content&view=article&id=12683:cricket-sl-armed-forces-to-maintain-three-test-venues&catid=58:news&Itemid=626). *Sunday Times (Sri Lanka)*. 5 November 2011. .
[48] Abeywickrema, Mandana Ismail (23 January 2011). "The Militarisation Of Sri Lanka's Diplomatic And Administrative Services" (http://www.thesundayleader.lk/2011/01/23/the-militarisation-of-sri-lankaâs-diplomatic-and-administrative-services/). *Sunday Leader*. .

External links

- Ministry of Defence, Public Security, Law & Order - Democratic Socialist Republic of Sri Lanka (http://www.defence.lk)
- The official web site of the Government of Sri Lanka (http://www.priu.gov.lk)
- Sri Lanka Army (http://www.army.lk)
- Sri Lanka Navy (http://www.navy.lk)
- Sri Lanka Air Force (http://www.airforce.lk)
- General Sir John Kotelawala Defence Academy (http://www.kda.lk/)
- Media Center for National Security (http://www.nationalsecurity.lk)
- The Permanent Mission of Sri Lanka to the United Nations Office at Geneva (http://www.lankamission.org)
- Library of Congress (http://lcweb2.loc.gov/cgi-bin/query/r?frd/cstdy:@field(DOCID+lk0155))
- Library of Congress Foreign Military Relations (http://lcweb2.loc.gov/cgi-bin/query/r?frd/cstdy:@field(DOCID+lk0172))
- Peace in Sri Lanka (http://www.peaceinsrilanka.org/peace2005/Insidepage/AtaGlance/Ceasefire_Feb-Dec2005.asp)
- Full-scale fighting flares in Sri Lanka (http://www.janes.com/defence/news/jdw/jdw060817_1_n.shtml) Extract from article about recent fighting, August 2006

List_of_Sri_Lanka_Army_regiments_and_corps

Sri Lanka Army

Components
Sri Lanka Army
Sri Lanka Army Volunteer Force
List of regiments and corps
Structure of the Sri Lanka Army
History
History of the Sri Lanka Army
Equipment
Weaponry and equipment
Personnel
Commander of the Army
Army Ranks

This is a list of all combat and combat support regiments and corps of the Sri Lanka Army.

Armoured

- Sri Lanka Armoured Corps

Infantry

Regular Army

- Sri Lanka Light Infantry
- Sri Lanka Sinha Regiment
- Gemunu Watch
- Gajaba Regiment
- Vijayabahu Infantry Regiment
- Mechanized Infantry Regiment
- President's Guard

Volunteer Force

- Sri Lanka Rifle Corps
- Sri Lanka National Guard

Special Forces

- Commando Regiment
- Special Forces Regiment

Support Arms and Services

Support Arms

- Sri Lanka Artillery
- Sri Lanka Engineers
- Sri Lanka Signals Corps

Services

- Military Intelligence Corps
- Engineer Services Regiment
- Sri Lanka Army Service Corps
- Sri Lanka Army Medical Corps
- Sri Lanka Army Ordnance Corps
- Sri Lanka Electrical and Mechanical Engineers
- Sri Lanka Corps of Military Police
- Sri Lanka Army General Service Corps
- Sri Lanka Army Women's Corps
- Sri Lanka Army Pioneer Corps

Disbanded

- Ceylon Rifle Regiment
- Ceylon Mounted Rifles
- Ceylon Planters Rifle Corps
- Colombo Town Guard
- Rajarata Rifles
- Ruhunu Regiment
- National Service Regiment
- Post and Telegraph Signals
- Ceylon Railway Engineer Corps

Sri Lanka Army

Sri Lanka Army	
 Sri Lanka Army Insignia and Flag	
Active	1 April 1881 - Present [1]
Country	Sri Lanka
Branch	Sri Lanka Armed Forces
Type	Army
Role	Foreign and Domestic Defense
Size	276,700[2]
Part of	Ministry of Defence
Military Headquarters	Army Headquarters, Colombo
Motto	Latin: *Pro Patria* "For Country"
Anniversaries	October 9
Engagements	World War II 1971 JVP Insurrection 1987–89 JVP Insurrection Sri Lankan Civil War
Commanders	
Commander of the Army	Lt.Gen. Jagath Jayasuriya
Notable commanders	Gen. D.S. Attygalle Gen. Sarath Fonseka Lt. Gen. Denzil Kobbekaduwa Maj. Gen. Vijaya Wimalaratne Maj. Gen. Bertram Heyn Maj. Gen. Anton Muttukumaru

Sri Lanka Army

Components

Sri Lanka Army

Sri Lanka Army Volunteer Force

List of regiments and corps

Structure of the Sri Lanka Army

History

History of the Sri Lanka Army

Equipment

Weaponry and equipment

Personnel

Commander of the Army

Army Ranks

Sri Lankan Civil War
(1983–2009)

Background

Sri Lanka · History of Sri Lanka

Origins of the Civil War

Origins of the Civil War · Sri Lankan Tamil nationalism · Sinhalese Buddhist nationalism · Riots · Black July

Main phases

Eelam War I · Indian intervention · Eelam War II · Eelam War III · Eelam War IV

LTTE

LTTE (Divisions) · Black Tigers · Attacks · Expulsion of Muslims · Suicide bombings

Military of Sri Lanka

Military · Civilian attacks · Army · Navy · Air Force · Police · Home Guards · LRRP · STF

Major leaders

M. Rajapaksa · V. Prabhakaran · C. Kumaratunga · A. Balasingham · J. R. Jayewardene · Karuna

Indian involvement

Operation Poomalai · Indo-Sri Lanka Accord · Indian Peace Keeping Force · Operation Pawan · Rajiv Gandhi · RAW

Militant & paramilitary groups

Tamil militant groups (List) · ENDLF · ENLF · EPDP · EPRLF · EROS · PLOTE · TELO · TMVP

Other

Battles · Casualties · War crimes · LLRC · State terror · Human rights · Disappearances · Child soldiers · Assassinations · Protests (Canada)

The **Sri Lanka Army** is the oldest and largest of the Sri Lanka Armed Forces and is responsible for land-based military and humanitarian operations. Established as the **Ceylon Army** in 1949, it was renamed when Sri Lanka became a republic in 1972. As of the year 2010, the Army has believed to have approximately 200,000 regular personnel, approximately between 20,000-40,000 reserve personnel and 18,000 National Guardsmen[3] [4] and comprises 13 operational divisions, one air-mobile brigade, one commando brigade, one special forces brigade, one independent armored brigade, three mechanized infantry brigades and over 40 infantry brigades.[5] From the 1980s to 2009 the army was engaged in the Sri Lankan civil war.

The professional head of the Sri Lanka Army is the Commander of the Army, currently Lieutenant General Jagath Jayasuriya.[6] The Commander-in-Chief of the Sri Lankan Military is the President of the country, who heads the National Security Council through the Ministry of Defense, which is the highest level of military command charged with formulating, executing defence policy and procurements for the armed forces.[7] However operations of the Sri Lanka Army are coordinated by the Joint Operations Command, with other two armed forces.

The Sri Lankan Army has also provided special training to the United States Army on their request as well as many other countries in military education regarding civilian rescue, jungle combat, and guerilla warfare etc.[8]

History

Ancient and pre-colonial times

The first military engagements in Sri Lankan history were marked by the advent of King Vijaya, a Bengal prince who landed along with his followers on the beaches of northwestern Sri Lanka around 543 BC. Prince Vijaya and his followers occupied the lands of the native Veddah people. Repeated incursions by South Indians, particularly the Cholas, into Sri Lankan territory occurred throughout the next few centuries and led to the engagement of the rival forces in battle.[9] In one famous encounter, Sinhalese King Dutugemunu (200 BC) raised an army of eleven thousand inhabitants in his battle against the Chola invader King Elara, whom he eventually defeated. King Dutugemunu's organizational skills, bravery and chivalry are famous and his battles have gone down in history as outstanding offensive operations.[10]

Other Sri Lankan rulers whose military achievements stand out include King Gajabâhu (113 AD), who sailed to India to bring back his captured soldiers, and King Dhatusena (433 AD) who is credited with repulsing numerous Indian invasions and for organizing a naval build-up to deter seaborne attacks. He also had the foresight to cover his defenses with artillery. Vijayabâhu I (1001 AD) was another warrior king who dislodged Indian invaders and united the country. Parakramabahu the Great (1153 AD) was an outstanding monarch of the Polonnaruwa period of Sri Lankan history, and his accomplishments as a military leader and a great administrator are noteworthy. His reign included a military expedition to Burma (Myanmar) in retaliation for indignities inflicted on his envoys and Burmese interference in the elephant trade. This marked the first overseas expedition in Sri Lankan military history. It is also reported that Parakramabahu's fame was such that his assistance was sought by South Indian rulers who were involved in internecine struggles. Another strong ruler in the pre-colonial era was Parâkramabâhu VI, who defeated Indian invaders, united the island and ruled it from capital Sri Jayawardhanapura, Kotte.[10] Although the known epigraphical records do not indicate that the Sri Lankan rulers had a full-time standing army at their disposal, there is evidence supported by legend, designation, name, place and tradition that prove there were 'stand-by' equestrian, elephant, and infantry divisions to ensure royal authority at all times. Militias were raised as the necessity arose, and the soldiers returned to their pursuits, mainly for farming, after their spell of military duty.[10]

Colonial era

Parts of Sri Lanka came under the control of three colonial European powers, namely the Portuguese in the 16th century, the Dutch in the 17th century and the British in the 18th century. Yet, until the entire island was ceded to the British in 1815, regional kingdoms maintained most of their independent defense forces and were able to successfully repulse repeated thrusts by the European armies. However the British, unlike their counterparts, were not primarily restricted to maritime power, and thus had the capability to bring the entire island under their control and to integrate locals into the British defense forces.[10]

Portuguese and Dutch rule (1505-1796 AD)

In the beginning of the 16th century, modern Europe first came in contact with Sri Lanka, which was then referred to as Ceylon. In 1505 a Portuguese fleet, while operating in the Indian seas against Arab traders, was blown off course and landed at Galle, on the southern coast of the island.[11] In 1517 the Portuguese re-appeared, and with the consent of the Sinhalese King established a trading post in Colombo. Having initiated contact with Sri Lanka as traders, the Portuguese soon made themselves political masters of the western seaboard. Numerous forts were soon established, and features of European civilization were introduced.[10]

The Portuguese are credited with the introduction of European-style fortresses to Sri Lanka during this era. Although some locals already possessed military training and fighting experience, there is no evidence that the Portuguese employed local inhabitants into their own forces. Thus the Portuguese were forced to restrict their presence in the island due to their small numbers and their efforts were more focussed toward projecting maritime power.[10]

In 1602 Dutch explorers first landed in Sri Lanka, which was then under Portuguese control. By 1658 they had completely ousted the Portuguese from the coastal regions of the island. Much like the Portuguese, they did not employ locals in their military, and preferred to live in isolation, pursuing their interests in trade and commerce. Like the Portuguese, they defended their forts with their own forces, but unlike the Portuguese, Dutch forces employed Swiss and Malay mercenaries. The Dutch Forts in Jaffna, Galle, Matara, Batticaloa and Trincomalee were sturdily built and are considered a tribute to their military engineering skills. Also, like the Portuguese, the Dutch focussed on maritime power and although they had the capability to develop and use local forces, they chose to isolate themselves from the local population.[10]

British rule (1798-1948 AD)

The British Empire then ousted the Dutch from the coastal areas of the country, and sought to conquer the independent Kandyan Kingdom. In the face of repeated British assaults, the Kandyans were forced into a degree of guerrilla warfare and fared well against their superior British adversaries.[10]

Initially the British stationed their forces, which included naval vessels, artillery troops and infantry, to defend of the island nation from other foreign powers, using the natural harbor of Trincomalee as their headquarters in Sri Lanka. In 1796, the Swiss and Malay mercenaries who were previously in the service of the Dutch were transferred to the British East India Company. While the Swiss De Meuron's Regiment was eventually disbanded in Canada in 1822, the Malays, who initially formed a Malay Corps, were converted into the 1st Ceylon Regiment in 1802 and placed under a British commanding officer. In the same year, the British became the first foreign power to raise a Sinhalese unit, which was named the 2nd Ceylon Regiment, also known as the Sepoy Corps.[10]

In 1803 the 3rd Ceylon Regiment was created with Moluccans and recruits from Penang. All these regiments fought alongside British troops in the Kandyan wars which began in 1803. Throughout the following years, more Sinhalese and Malays were recruited to these regiments, and in 1814 the 4th Regiment was raised, which was composed entirely of African troops. It was later renamed as the Ceylon Rifle Regiment. Eventually, the Kandyan Kingdom was ceded to the British in 1815, and with that they gained control over the whole island. Resistance to British occupation cropped up almost instantly. During the first half-century of occupation, the British faced a number of uprisings, and were forced to maintain a sizable army in order to guarantee their control over the island. After the Matale Rebellion led by Puran Appu in 1848, in which a number of Sinhalese recruits defected to the side of the rebels, the recruitment of Sinhalese to the British forces was temporarily halted.[10]

Ceylon Light Infantry Volunteers

The second phase in the employment of non-British personnel commenced in 1861 after the enactment of an ordinance designed to authorize the creation of a Volunteer Corps in the island. It was designated the Ceylon Light Infantry Volunteers (CLIV). This move compensated for the disbandment of the Ceylon Rifle Regiment in 1874. The Ceylon Light Infantry Volunteers was originally administered as a single unit. However, over the years various sections of the volunteers grew large enough to become independent from their parent unit. The different units that emerged from the Volunteer Force were the

- Ceylon Artillery Volunteers
- Ceylon Mounted Infantry (CMI)
- Ceylon Volunteer Medical Corps
- Cadet Battalion Ceylon Light Infantry
- Ceylon Engineers
- Ceylon Supply & Transport Corps
- Ceylon Planters Rifle Corps (CPRC).[10]

Ceylon Defence Force

In 1910 the name of the military was formally changed to the Ceylon Defence Force (CDF). It continued to grow throughout the early period of the 20th century. The CDF saw active service when a contingent of the Ceylon Mounted Infantry (CMI) in 1900, and a contingent of the Ceylon Planters Rifle Corps (CPRC) in 1902, took part in the Second Boer War in South Africa. Their services were recognized by the presentation in 1902 of a color to the CMI, and a presentation in 1904 of a banner to the CPRC. In 1922, the CDF was honored by the presentation of the King's and Regimental colors to the Ceylon Light Infantry (CLI).[10]

First Prime Minister of Independent Sri Lanka Hon. D.S.Senanayaka visiting the 1st battalion of the CLI at the Echelon Square and watching volunteers being trained to handle light machine guns.

During the First World War, many volunteers from the Defence Force traveled to England and joined the British Army, and many of them were killed in action. One of them mentioned by Sir Arthur Conan Doyle was Private Jacotine of the CLI, who was the last man left alive in his unit at the Battle of Lys, and who continued to fight for 20 minutes before he was killed.[12]

In 1939, the CDF was mobilized and an enormous expansion took place which required the raising of new units such as the Ceylon Signals Corps, the Auxiliary Territorial Service (Ceylon) and also the Colombo Town Guard, which had been previously disbanded, but was later re-formed to meet military requirements. During the Second World War, Britain assumed direct control over the Armed Forces of Ceylon.[13]

Post-independence

Brigadier James Sinclair, Earl of Caithness inspecting a guard of honour wearing khaki drill.

At the end of World War II, CDF which had increased in size during the war began demobilization.In 1948 Sri Lanka gained independence from Britain, becoming a Dominion within the commonwealth and a year earlier Ceylon entered into the bi-lateral Anglo-Ceylonese Defence Agreement of 1947. This followed by the Army Act No. 17 of 1949 which was passed by Parliament on April 11, 1949 and formalized in Gazette Extraordinary No. 10028 of October 10, 1949 marked the creation of the Ceylon Army, consisting of a regular and a volunteer force, the later being the successor of the disbanded CDF.[14] [15] Therefor October 10, 1949 is considered the Ceylon Army was raised, and October 10 is celebrated annually as Army day. The Defence Agreement of 1947 provided the assurance that British would come to the aid of Ceylon in the event it was attacked by a foreign power and provided British military advises to build the country's military. Brigadier James Sinclair, Earl of Caithness was appointed as general officer commanding Ceylon Army, as such becoming the first commander of the Ceylon Army.

The initial requirement was to raise an artillery regiment, an engineer squadron, an infantry battalion, a medical unit, and a service corps company. For much of the 1950s the army was preoccupied with the task of building itself and training existing and new personal. To this aim the British Army Training Team (BATT) advisory group carried out training for ex-members of the CDF within the Ceylon Army, senior officers were sent to the British Army Staff College, Camberley and some attached to units of the British Army of the Rhine to gain field experience. New officers were sent for training at Royal Military Academy, Sandhurst which continued until the 1960s and both officers and other ranks were sent to specialist training courses in Britain, India, Pakistan and Malaya. There were no formations and all units were structured to directly function under the Army Headquarters. However temporary field headquarters were to be formed at the time requirement arose.[15]

Due to a lack of any major external threats the growth of the army was slow, and the primary duties of the army quickly moved towards internal security by the mid 1950s, the same time as the first Ceylonese Commander Major General Anton Muttukumaru took command of the army. The first internal security operation of the Ceylon Army began in 1952, code named *Operation Monty* to counter the influx of illegal South Indian immigrants brought in by smugglers on the north-western cost, in support of Royal Ceylon Navy coastal patrols and police operations. This was expanded and renamed as Task Force Anti-Illicit Immigration (TaFII) in 1963 and continued up to 1981 when it was disbanded. The Army was mobilized to help the police to restore peace under provincial emergency regulations during the 1953 hartal, the 1956 Gal Oya Valley riots and in 1958 it was deployed for the first time under emergency regulations throughout the island during the 1958 Riots.[16]

During the 1950s and 1960s the army was called apron to carry to essential services when the workers went on strike which were organized by the left wing parties and trade unions for various reasons, the most notable was the 1961 Colombo Port strike, during which ships threatened to bypass Colombo port and the country almost staved. To counter these common strikes several units were formed, who were employed in development work when there were no strikes.[16]

In 1962 several volunteer officers attempted a military coup, which was stopped hours before it was launched. This attempted coup effect the military to a great extent, since the government mistrusted the military, it reduced the size and growth of the army, especially the volunteer force, with several units being disbanded. In May 1972, when Ceylon was proclaimed a republic and changed its name to from the Dominion of Ceylon to the Republic of Sri Lanka, all Army units were renamed accordingly.[17]

1970–Present

After successfully defeating the insurgency led by the Janatha Vimukthi Peramuna (JVP) in 1971, the army was confronted with a new conflict, this time with the Liberation Tigers of Tamil Eelam (LTTE) and other Tamil militant groups. The war escalated to the point where India was asked to intervene as a peacekeeping force. This was later seen as a tactical error, as the Indian Peace Keeping Force united nationalist elements such as the JVP to politically support the LTTE in their call to evict the IPKF. The war with the LTTE was halted following the signing of a ceasefire agreement in 2002 with the help of international mediation. However, renewed violence broke out in December 2005 and following the collapse of peace talks, the Army has been involved in the heavy fighting that has resumed in the north and east of the country. Since 1980 the army has undertaken many operations against the LTTE rebels. The major operations conducted by the army eventually led to the recapture of Jaffna and other rebel strongholds. In 19 May 2009 Sri Lankan army declare the victory of war as they found the dead body of LTTE leader Velupillai Prabhakaran. This marked the end of the war, with the LTTE ceasing to exist in Sri Lanka as a result of prolonged military offensives conducted by Sri Lanka army.[18]

Major combat operations

- First JVP Insurrection (1971–1972)
- Eelam War I (1976–1987)
  - Operation Liberation
- Second JVP Insurrection (1987–1990)
- Eelam War II (1990–1995)
  - Operation Sea Breeze
  - Operation Thrividha Balaya
  - Operation Balavegaya I, II
- Eelam War III (1995–2002)
  - Operation Riviresa
  - Operation Jayasikuru
  - Operation Rivibala
  - Operation Ranagosa
  - Operation Rivikirana
  - Operation Kinihira I, II, III/IV, V/VI, VII, VIII ,IX
- Eelam War IV (2006–2009)
  - Eastern Theater
    - Operations in Thoppigala
  - Northern Theater
    - Battle of the Forward Defence Lines

Peacekeeping

The Sri Lanka Army has taken part in two peacekeeping missions with United Nations over the course of its history. First assignment was in the Congo (ONUC) (1960–1963). Most recently, following the signing of a ceasefire agreement was signed between the government and the LTTE in 2002, Sri Lankan forces were invited by the United Nations to be part of the UN peacekeeping force in Haiti.[19] The Sri Lanka Light Infantry was selected to be part of the peacekeeping operations, and training programs on counter revolutionary warfare were conducted for the troops in Kukuleganga. Following the successful completion of the training, a battalion under the command of Col. K A D A Karunasekara left for Haiti on October 22, 2004. In the process of the peacekeeping operations, 2 soldiers were killed in a raid in Petit-Goave.[20] After over 6 months of service, the first contingent of the peacekeeping force returned to Sri Lanka on May 17, 2005.[19] In December 2007, 7th rotation of the Sri Lankan contingent had been deployed with a force of 991 officers and other ranks, many of the those deployed have been awarded the United Nations Medal for their services.[21]

Peacekeeping Sex Scandal

In November 2007, 114[22] members of the 950 member Sri Lankan Army peacekeeping mission in , were accused of sexual misconduct and abuse.[23] [24] 108 members, including 3 officers of the 950-member-strong Sri Lanka peacekeeping contingent is being sent back after being implicated in alleged misconduct and sexual abuse.[25] UN spokeswoman Michele Montas said: "The United Nations and the Sri Lankan government deeply regret any sexual exploitation and abuse that has occurred."[24] The Sri Lankan Officials claim that there is little tangible evidence on this case.[22] After inquiry into the case the UN Office of Internal Oversight Services (OIOS) has concluded 'acts of sexual exploitation and abuse (against children) were frequent and occurred usually at night, and at virtually every location where the contingent personnel were deployed.' The OIOS is assisting in the pending legal proceedings initiated by the Sri Lankan Government and has said charges should include rape "because it involves children under

18 years of age".[26]

Current deployments

As of present, the bulk of the Sri Lankan Army is deployed for domestic defensive and combat operations, while a sizable foreign deployment is maintained.

Domestic

Due to the Sri Lankan Civil War the army has been on a constant mobilized (including reservist) state since the 1980s (except for a brief period from 2002–2005). The majority of the army as been deployed in the North and Eastern provinces of the country, which includes 14 Divisions coming under six operational headquarters and 2 independent Divisions and several independent Brigades. The army is also based in other parts of the island for internal security including a Division for the defense of the capital.

Foreign

The Sri Lanka Army currently participates in several major overseas deployments:

- Haiti - an infantry battalion with support personal totaling around 1000 personal in Haiti as part of the United Nations Stabilization Mission in Haiti since 2004.[27]
- Chad - a contingent of engineers joined the United Nations Mission in the Central African Republic and Chad on May 25, 2010.[28] [29]
- Lebanon - a mechanized infantry company with combat support personal in the United Nations Interim Force in Lebanon since November 2010.[30]

Organization Structure

The professional head of the army is the Commander of the Army, at present Lieutenant General Jagath Jayasuriya. He is assisted by the Chief of Staff of the Army, currently Major General Daya Ratnayaka The Commandant of the Volunteer Force is head of the Army Volunteer Force and is responsible for the administration and recruitment of all reserve units and personal. The Army Headquarters, situated in Colombo is the main administrative and the operational headquarters of the Sri Lanka Army.

Administrative

The Army Headquarters is divided into a number of branches, namely the General Staff (GS) branch responsible for coordination of operations and training and the Adjutant General's (AGs) branch responsible for personal administration, welfare, medical services and rehabilitation. The Quarter Master General's (QMGs) branch is responsible for feeding, transport, movement and construction and maintenance. The Master General of Ordnance's (MGOs) branch is responsible for procurement and maintenance of vehicles and special equipment.[31] The Military Secretary's Branch is responsible for handling all matters pertaining to officers such as promotions, postings and discipline. Each branch is headed by an officer in the rank of Major General who is directly responsible to the Commander of the Army for the smooth functioning of the Branch. Under each Branch, there are several Directorates, each headed by a Brigadier.[31]

The Headquarters of field formations each have its own staff. For instance a divisional headquarters is divided into a GS branch as an AQ branch, each headed by a Colonel and is responsible for operations & training and administration & logistics respectively. Similarly, a Brigade Major and Major AQ is responsible for operations and administration in a brigade.[31]

Like the Indian Army, the Sri Lanka Army has largely retained the British-style regimental system that it inherited upon independence. The individual regiments (such as the Sri Lanka Light Infantry and the Sinha Regiment) operate independently and recruit their own members. Officers tend to remain in a single battalion throughout their careers. The infantry battalion, the basic unit of organization in field operations, includes five companies of four platoons each. Typical platoon have three squads (sections) of ten personnel each. In addition to the basic infantry forces, a commando regiment was also established in 1986. Support for the infantry is provided by an armoured regiment, five reconnaissance regiments, three mechanized infantry regiments, five field artillery regiments, a rocket artillery regiment, three commando regiments , three special forces regiments, six field engineering regiments, five signals battalions, a medical corps, and a variety of logistics units.[32]

Regiments & Corps

[14]

Name	Headquarters	Subunits
Armoured Corps	Rock House Army Camp, Colombo	Eight regular regiments and a volunteer regiment.
Artillery	Panagoda Cantonment, Panagoda	Seven regular regiments and two volunteer regiments.
Engineers	Army Headquarters, Colombo	Six regular regiments and one volunteer regiment.
Signals Corps	Panagoda Cantonment, Panagoda	Seven regular regiments and one volunteer regiment with Signals workshop and Infonation Technologe Uint (11 SLSC)
Light Infantry	Panagoda Cantonment, Panagoda	Ten regular battalions, five volunteer battalions.
Sinha Regiment	Ambepussa Camp, Ambepussa	Seven regular battalions, five volunteer battalions and a head quarter battalion.
Gemunu Watch	Kuruwitha Army camp, Ratnapura	Nine regular units, four volunteer units.
Gajaba Regiment	Saliyapura Camp, Anuradhapura	Twelve regular battalions and five volunteer battalions.
Vijayabahu Infantry Regiment	Boyagane Camp, Kurunegala	Eight regular battalions and four volunteer battalions.
Mechanized Infantry Regiment	N/A	Three regular battalions.
President's Guard	N/A	N/A
Commando Regiment	Ganemulla, Gampaha	Four regular regiments.
Special Forces Regiment	Seeduwa, Gampaha	Three regular regiments.
Military Intelligence Corps	Colombo	Two regular battalions.
Engineer Services Regiment	Panagoda Cantonment, Panagoda	Three regular regiments and a volunteer regiment.
Service Corps	Panagoda Cantonment, Panagoda	Four regular units and one volunteer unit.
Medical Corps	Panagoda Cantonment, Panagoda	Two regular units and one volunteer unit.
Ordnance Corps	Panagoda Cantonment, Panagoda	Three regular ordnance battalions and one volunteer ordnance battalion.
Electrical and Mechanical Engineers	Slave Island, Colombo	Three regular regiments and one volunteer regiment.
Corps of Military Police	Polhengoda, Colombo	Three regular regiments.

General Service Corps	Panagoda Cantonment, Panagoda	Three regular unit and a volunteer unit.(With 3 & 4 SLAGSC (Pay & Recored)
Women's Corps	Regtl Centre, Borella	One regular unit and 4 volunteers unit.
Rifle Corps	Army Headquarters, Colombo	Two volunteer battalions.
Pioneer Corps	Headquarters, Battharamulla, Pelawattha.	One volunteer unit.
National Guard	Kurunegala	32 volunteer battalions.

Operational Command

Organized and controlled by the Army General Staff at Army HQ, various formations are raised from time to time to suit various security requirements and operation in the country and overs seas. The Army at present has deployed 12 Divisions, 7 task forces and several independent brigades. Except for the 11 Division based at the Panagoda Cantonment which is responsible for the maintenance of capability for the defence of the capital, all other divisions, task forces and brigades are deployed for operations in the Northern and Eastern provinces of Sri Lanka, coming under six regional commands known as *Security Forces Headquarters*, which are the Security Forces Headquarters Jaffna (SFHQ-J), Wanni (SFHQ-W), East (SFHQ-E), Kilinochchi (SFHQ-KLN), Mullaittivu (SFHQ-MLT) & South (SFHQ-S)

Each SFHQ and most divisions are commanded by a General Officer Commanding in the rank of Major General. A SFHQ has several divisions under its command and each division is further divided into brigades. Each brigade is commanded by an officer in the rank of Brigadier and has a number of Infantry battalions, support arms (Artillery, Engineers and Signals) and support services (Service Corps, Engineering Services, Ordnance Corps, Electrical and Mechanical Engineers) under assigned to it. There are also several independent brigade (Air Mobile Brigade, Armored Brigade, etc.)

In other parts of the country, there are Area and Sub-Area Headquarters. Armour, Artillery, Engineers and Signals Units are grouped under Brigade Headquarters of their own arm; Armored Brigade, Artillery Brigade and so on.

Force formations

[33]

Security Forces Headquarters - Jaffna (SFHQ-J)

- 51 Division, based in Jaffna
- 52 Division, based in the Jaffna Peninsula
- 55 Division, based in Elephant Pass Military Base, Jaffna Peninsula [33]

Security Forces Headquarters - Wanni (SFHQ-W) [33]

- 56 Division, operating in the Vavuniya District
- 61 Division, operating in the Vavuniya District
- 21 Division
- Area Headquarters Mannar, Mannar

Security Forces Headquarters - East (SFHQ-E)

- 22 Division, based in Trincomalee[34]
- 23 Division, based in Poonani, Batticaloa District [35]

Security Forces Headquarters - Kilinochchi (SFHQ-KLN)[33] [36]

- 57 Division, operating in the Kilinochchi District
- 66 Division, operating in the Kilinochchi District
- 68 Division, operating in the Kilinochchi District

- Task Force 3, operating in the Kilinochchi District
- Task Force 7, operating in the Kilinochchi District

Security Forces Headquarters - Mullaittivu (SFHQ-MLT) [33]

- 59 Division, operating in the Mullaittivu District
- 64 Division, operating in the Mullaittivu District
- 65 Division, Thunukkai, Mullaittivu District
- Task Force 2, operating in the Mullaittivu District

Security Forces Headquarters - South (SFHQ-S)[37] [38]

- Operation Command Colombo, Colombo
- 11 Division, based at the Panagoda Cantonment, Western Province
  - 111 "Kandy" Brigade
- Area Headquarters Hambantota, Hambantota
- Area Headquarters Galle, Galle
- Area Headquarters Diyatalawa, Diyatalawa
- Sub Area Headquarters Ratnapura, Ratnapura
- Sub Area Headquarters Kurunegala, Kurunegala

Independent Divisions

- 53 Division, based at Mankulam [33]
- 58 Division, based at Paranthan (formally referred to as the Task Force 1) [33]

Independent Brigades

- Air Mobile Brigade
- Mechanized Infantry Brigade
- Armored Brigade
- Special Forces Brigade
- Commando Brigade
- Artillery Brigade
- Engineer Brigade
- Signals Brigade

Disbanded

- 54 Division, based at Elephant Pass[39]
- 2 Division
- 3 Division

Training

General Sir John Kotelawala Defence University (KDU) formed in 1981 and situated in Ratmalana, fourteen kilometers south of Colombo, is Sri Lanka's only university specializing in defense studies. Each year, approximately fifty cadets from all three services are admitted to the university (aged 18–22) to participate in a three-year program of academic work and basic training.[31] [32]

Junior field officers of the army and their counterparts in the Navy and Air Force are given advanced training and education at the **Defence Services Command and Staff College** (DSCSC) at Batalanda, Makola which was established in 1997 as the *Army Command and Staff College.*

Basic officer training is carried out at the **Sri Lanka Military Academy** (SLMA) (formally the **Army Training Center**) situated in Diyatalawa, in the Badulla District. The officer cadets graduating from the academy are commissioned as officers in the regular and volunteer forces. The course for officer cadets runs for ninety weeks and includes training in tactics and administration which helps prepare the cadets to take up the positions of platoon commanders. The course consisted of military and academic subjects and also trained the cadets physically. The course helps to promote leadership qualities and the understanding of each one's role as an officer and a servant of the state. Due to the lack of officers within the lower levels, the training process was sped up in the 1980s by developing a short commission course. The cadets were given a training of fifty-six weeks and devoted themselves to continue their careers in the military with the ten years of service for regular army officers and five years of service for volunteer officers.

Training for the new recruits are carried out by the **Army Training School** in Maduru Oya at several locations followed by additional training (both officers and other ranks) at the **Infantry Training Centre** in Minneriya, the **Combat Training School** in Ampara, while non-commissioned officers receive training at the **Non-commissioned Officers Training School** at Kala Oya. All these establishments come under the control of the Directorate of Training, Army Headquarters. Specialist and additional training is given by specialist training schools, regimental training centers and individual field units.[31] [32]

As the armed forces of Sri Lanka have a limited indigenous training facilities, especially in advanced roles, they have depended greatly on military training provided by foreign countries. The United Kingdom played a major role in the early years following independence and have continued to be an important source of military expertise to the Sri Lankan military. Other sources include Pakistan, Australia, Malaysia, and the United States. Additionally, in an agreement reached in 1984, Israeli security personnel (reportedly from Shin Bet, the Israeli counterespionage and internal security organization) trained army officers in counterinsurgency techniques.[32]

The Sri Lankan Army has also provided special training to the United States Army on their request as well as many other countries in military education regarding civilian rescue, jungle combat, and guerilla warfare etc.[8]

Training Centers[40]

1. Sri Lanka Military Academy (SLMA)
2. Army Training School (ATS)
3. Infantry Training Centre (ITC)
4. Combat Training School (CTS)
5. Army Physical Education Centre (APEC)
6. Volunteer Force Training School (VFTS)
7. Marksman Sniper Training School (MSTS)
8. Non-Commission Officer Training School (NCOTS)
9. Language Training School (LTS)
10. Institute of Peacekeeping Support Operations Training Sri Lanka (IPSOT-SL)

Regimental Training Centers[40]

1. Armoured Cops Training Centre
2. School Of Artillery
3. Sri Lanka School of Military Engineering
4. School Of Mechanical Engineers
5. School Of Signals
6. Commando Regiment Training School
7. Engineer Service School
8. Sri Lanka Army Service Corps. School
9. Sri Lanka Army Military School Of Nursing
10. Sri Lanka Army Ordnance School
11. Sri Lanka Electrical And Mechnaical Engineers School
12. Sri Lanka Corps of Military Police School
13. SriLanka Army General Service Corps. School

Personnel

The Sri Lanka Army presently stands at 200,000 strong[2] including 2,960 women plus and 58,000 reservists.[41]

In late 1987, the army had a total estimated strength of up to 40,000 troops, about evenly divided between regular army personnel and reservists on active duty. The approximately 20,000 regular army troops represented a significant increase over the 1983 strength of only 12,000. Aggressive recruitment campaigns following the 1983 riots raised this number to 16,000 by early 1985.[32] By 2007, the Army had expanded to over 120,000 personnel.[3]

Since the Sri Lankan armed forces are all volunteer services, all personal in the Sri Lanka Army have volunteered as regular personnel or reservists. This should not be confused with the traditional term *volunteers* used for reservists or reservist units. Recruitment of the personal are carried island wide with a restrictions in the norther and eastern provinces during the civil war in those areas. The Rifle Corps is the only territorial unit that carries out recruitment from a specific area. In June 2009, Sri Lanka announced plans to create a "Tamil regiment" to promote integration in the army[42]

Parama Weera Vibhushanaya recipients

The Parama Weera Vibhushanaya is the highest award for valour awarded in the Sri Lankan armed forces. Army recipients include;

- Colonel A.F. Lafir †
- Major G. S. Jayanath †
- Captain Saliya Upul Aladeniya †
- Second Lieutenant K.W.T. Nissanka †
- Warrant Officer 2nd Class Pasan Gunasekera †
- Corporal Gamini Kularatne †
- Lance Corporal W. I. M. Seneviratne †

Notable fallen members

Over 23,790 Sri Lankan armed forces personnel were killed since begin of the civil war in 1981 to its end in 2009, this includes 11 general officers killed in active duty or assassinated.[43] 659 service personnel were killed due to the second JVP insurrection from 1987 to 1990. 53 service personnel were killed and 323 were wounded in the first JVP insurrection from 1971 to 1972.[44] Notable fallen members includes;

- Lt. General Denzil Kobbekaduwa † - One of the greatest generals in modern Sri Lanka and Overall Operations Commander, Northern Sector.[45]
- Lt. General Parami Kulatunga † - Former Deputy Chief of Staff of the Army.[45]
- Lt. General Nalin Angammana † - Former GOC 3 Division.[46]
- Maj. General Vijaya Wimalaratne † - Jaffna Brigade Commander & One of the greatest generals in modern Sri Lanka.[45]
- Maj. General Percy Fernando † - Former deputy GOC 54 Division.[46]
- Maj. General Larry Wijeratne † - Former brigade commander, 51-4 Brigade.[46]
- Maj. General Lakshman 'Lucky' Wijayaratne † - Former brigade commander, 22 Brigade.[46]
- Maj. General Susantha Mendis † - Former brigade commander, 51-2 Brigade.[46]
- Maj. General Ananda Hamangoda † - Former brigade commander, 51-2 Brigade.[47]
- Brigadier Bhathiya Jayatilleka † - former Brigade commander, 54-1 brigade
- Brigadier Rohitha Neil Akmeemana † - former Brigade commander, Elephant pass.[46]
- Colonel Tuan Nizam Muthaliff - Former commanding officer 1st Battalion Military Intelligence Corps.[48]

Women in the Sri Lanka Army

The Sri Lanka Army Women's Corps (SLAWC) was formed on September 1, 1979 as an unarmed, noncombatant support unit. Set up with the assistance of the Women's Royal Army Corps, it was identical in structure to its parent organization, and its first generation of officer cadets was trained in Britain. Candidates were required to be between eighteen and twenty years old and to have passed the General Common Entrance (Ordinary level) examinations, while the Officer candidates must have passed the Advanced Level. Enlistment entailed a five-year service commitment (the same as for men), and recruits were not allowed to marry during this period. In the sixteen-week training course at the Army Training Center at the Diyatalawa Sri Lanka Military Academy, cadets were put through a program of drill and physical training similar to the men's program, with the exception of weapons and battle craft training. Female recruits were paid according to the same scale as the men, but were limited to service in nursing, communications, and clerical work. In late 1987, the first class of women graduates from the Viyanini Army Training Center were certified to serve as army instructors. But, from late 1987 - after hostilities began, the first batch of women graduates from the British Army's Women's Corp Center certified to serve as Army Instructors.[49]

Up to now, women officers have proved their ability and serve in varied specialized fields in the Service as control tower operators, electronic warfare technicians, radio material teletypists, automotive mechanics, aviation supply personnel, cryptographers, doctors, combat medic, lawyers, engineers and even aerial photographers.[50]

To meet the operational requirements in the field areas, the 2nd Volunteer Battalion of the Women's Corps was also raised. A few officers from the regular counter part were attached to this unit to organize the command structure. They are currently employed in active combat duties in the northern and eastern parts of the island.

Many officers commencing with Lieutenant Colonel A.W. Thambiraja were appointed to command this unit from time to time. The first women's corps officer to command the unit was Lieutenant Colonel Kumudini Weerasekara in 1992 and as of 2007 there were three female officers of the rank of Major General. At present there is one regular regiment and four volunteer regiments in the Women's Corps.[51]

Complaints over survivors' benefits

The Organisation for Disabled Soldiers has complained that compensation paid for war victims is inadequate. In August 2007, a spokesman for the organisation pointed out that the compensation paid to the families of soldiers killed during the war has remained flat for 23 years at Rs. 150,000, which has dwindled in value to the equivalent of US$1500. This amount is the same for all ranks, including generals. In contrast, however, the families of politicians are much more highly compensated; the family of the late Minister Mr. M. H. M. Ashroff was awarded Rs. 5 million (US$50,000) following his death in a helicopter crash.

Equipment

In the 1980s and 1990s, the army expanded its range of weapons from the original stock of World War II-era British Lee Enfield rifles, Sten Submachine guns, Vickers machine guns, Bren machine guns, 6-inch coastal guns, Daimler Armoured Cars, Bren Gun Carriers,[52] 40 mm anti-aircraft guns, 3.7 inch heavy anti-aircraft guns and 4.2-inch heavy mortars as well as post war Alvis Saladins, Alvis Saracen, Ferrets and Shorland S55s. New sources of weaponry in the mid-to-late 1970s included the Soviet Union, Yugoslavia, and China - countries with which the leftist Bandaranaike government had close ties. China continued to be an important source of arms well into the 1990s.[32]

To meet the threat posed by predominantly the LTTE, Army purchased modern military hardware including 50-caliber heavy machine guns, rocket-propelled grenade (RPG) launchers, Night Vision Devices, 106 mm recoilless rifles, 60 mm and 81 mm mortars, 40 mm grenade launchers and some sniper rifles. Refurbished armored personnel carriers were added to the 'A' vehicle fleet of the 1st Reece Regiment, Sri Lanka Armoured Corps. These APCs enabled the Armoured Corps to have their own assault troops to provide close contact protection to their Alvis Saladin and Ferret Scout Cars which were vulnerable to anti-tank weapons. The capability of the Sri Lanka Artillery was enhanced with the introduction of Ordnance QF 25 pounders.[32] [53] Chinese-made 122 mm, 130 mm and 152 mm howitzers were introduced to the Sri Lankan Army in 1995 and 1998 whilst 122 mm Multi Barrel Rocket Launchers (MBRL), were first used in 2000 by the Sri Lanka Army.[54]

Though the weapons were obsolete at the time of purchase, the Government security forces found them to be successful in combat. Land mines proved to be the most lethal threat to the security forces, as many mines have been deployed against government forces by the LTTE in the Northern and Eastern Provinces. These mines were deployed with normalcy against government trucks and buses, with a high rate of casualty. These land mines weighed approximately 50 – 100 kg, against which no armoured vehicle that the SLA possessed was able to withstand the blast effect. Therefore the South African made Buffel was introduced to service in 1985 to reduce damage due to land mines. By 1987 the Army's own Unicorn was also developed to a level so that they too matched the capabilities of the Buffels from South Africa, this was followed up by the newer Unibuffel class.[55] Both the Unicorn and the Unibuffel are assembled by the Sri Lanka Electrical & Mechanical Engineers.[32] [53]

In recent years, Sri Lanka has become increasingly reliant on China for weapons.[56] This is due to most European nations and the United States Governments passing regulations about the selling of weaponry to nations which are suffering from internal conflict.[57] However in light of recent attacks by the LTTE, the United States has expressed its intent to maintain military training assistance and possibly increase it should the violence continue.

China has no such regulations upon their arms producers, and some see the sales as an attempt to gain political influence with strategically important Sri Lanka.[58] Sri Lanka also continues to receive a variety of weapons from Britain, Pakistan, Israel and other former suppliers.[56] [59]

Armour

Type	Origin	Quantity	Notes
Main Battle Tanks			
T-55/T-55AM2[55]	Soviet Union	140+	
Type 69[60]	China	20+	
Type 59	China	80+	
Light Tanks			
Type 63	China	N/A	Amphibious
Infantry fighting vehicles			
BMP-3[61]	Soviet Union	45	
BMP-2	Soviet Union	240	
BMP-1	Soviet Union	13	
Armoured personnel carriers			
Norinco Type 89 (YW534)[62]	China	N/A	Tracked
Type 85 (YW531H)	China	N/A	Tracked, Amphibious
Type 63 (YW531)[63]	China	N/A	Tracked
BTR-80	Soviet Union	45	wheeled
Type 92 (WZ551)	China	200	wheeled
Unibuffel	Sri Lanka	33+	Locally manufactured, Mine-protected APC
Unicorn	Sri Lanka	60+	Locally manufactured, Mine-protected APC
Buffel	South Africa	N/A	Mine-protected APC
BTR-152	Soviet Union	N/A	wheeled, (retired)
Hotspur Hussar[31]	United Kingdom	N/A	wheeled, (retired)
Shorland S55	United Kingdom	N/A	wheeled, (retired)
Hotspur[31]	United Kingdom	N/A	wheeled, (retired)
Reconnaissance vehicles			
Alvis Saladin	United Kingdom	N/A	wheeled (retired)
Alvis Saracen	United Kingdom	N/A	wheeled, (retired)
Daimler Ferret	United Kingdom	N/A	wheeled, (retired)
Engineering Support Vehicles			

VT-55[64]	Soviet Union	16	Armoured recovery vehicle
MT-55A[65]	Soviet Union	8	Armoured vehicle-launched bridge

Artillery

Type	Origin	Quantity	Notes
Rocket artillery			
RM-70 Multiple rocket launcher[64]	Czechoslovakia	30+	122.4 mm Multiple rocket launcher
BM-21 Multiple rocket launcher[64]	Soviet Union	5	122.4 mm Multiple rocket launcher
Towed artillery			
Type 66 152 mm gun-howitzer	China	36	152 mm gun-howitzer
Type 59 130mm field gun	China	12	130 mm field gun
Type 60 122mm howitzer	China	N/A	122 mm howitzer
Ordnance QF 25 pounder	United Kingdom	N/A	field guns - Ceremonial Gun Troop
Ordnance QF 75 mm	United Kingdom	N/A	field guns - Ceremonial Gun Troop
Mortars			
Type 86 (W86)	China	55	120 mm towed mortar
Type 84 (W84)	China	N/A	82 mm mortar
Type 89	China	N/A	60 mm light mortar
Weapon Locating Radar			
AN/TPQ-36 Firefinder Radar	United States	N/A	Weapon Locating and Counter-battery Radar
SLC-2 Radar	China	N/A	Weapon Locating and Counter-battery Radar

Infantry weapons

Handguns
Beretta M9 Pistol
Glock 17
Enfield revolver
Assault Rifles
AK-47 Assault rifles
Sa 58 Assault rifles
Type 56 Assault rifles[66]
Type 81 Assault rifles[67]
Heckler & Koch G3 Assault rifles[68]
FN FNC Assault rifles (replaced the FN FAL as far back as in 1981)
M16 Assault rifles[69]

M4 Carbine
SAR-80 Assault rifles[66]
Type 95 Assault rifles
Sub-Machine guns
H&K MP5 Submachine Guns[66]
Uzi Submachine Guns[66]
Sterling submachine guns[66]

Sniper Rifles
Accuracy International L96A1 Sniper Rifles[70]
Heckler & Koch PSG1 Sniper Rifles
Dragunov Sniper Rifles[71]
Zastava M93 Black Arrow[72]
Machine guns
Type 80 GPMG (Chinese version of Russian PKM)[73]
FN Minimi General purpose machine gun[66]
FN MAG General purpose machine gun[66]
Type 56 LMG (Chinese version of Russian RPD)[32]
HK21 Belt-fed light machine gun[74]
Grenade launchers
Milkor MGL Grenade launcher[66]
HK 69 Breech-loading grenade launcher[66]
M203 Grenade launcher[66]
Rocket launchers
RPO-A Shmel man-portable rocket launcher
Type 69 RPG rocket launchers (Chinese version of RPG-7)[32]
Anti-tank missiles
Baktar-Shikan Anti-tank guided missile

See also

- Awards and decorations of the military of Sri Lanka
- Sri Lanka Military Academy
- Sri Lankan Civil War
- [[Rehabilitation Directorat, Sri Lanka Army

Further reading

- Army, Sri Lanka. (1st Edition - October 1999). *Sri Lanka army: 50 years on, 1949-1999* ISBN 978-955-8089-02-6

References

[1] http://en.wikipedia.org/wiki/Sri_Lanka_Light_Infantry

[2] http://www.sundayobserver.lk/2010/07/25/sec04.asp

[3] "Country Reports on Human Rights Practices - 2000" (http://www.state.gov/g/drl/rls/hrrpt/2000/sa/index.cfm?docid=704). U.S. State Department. February 23, 2001. . Retrieved 2007-03-23.

[4] Army now 150,000 strong recruits more, readies for big battles on the Vanni front. The Island (http://www.island.lk/2008/02/11/news2.html)

[5] "Deadly arsenals dot Sri Lanka" (http://www.atimes.com/atimes/South_Asia/HH05Df02.html). *Richard M Bennett* (Asia Times Online). 5 August 2006. .

[6] Army, Navy get new commanders (http://www.island.lk/2009/07/13/news20.html)

[7] "Commander-in-Chief Completes One Year in Office" (http://web.archive.org/web/20070928111958/http://www.rnhit.com/mcns/articles/commander-in-chief.html?PHPSESSID=8a9729bc19d560d81319bc8bee3f380a). Media Center for National Security. 2007-03-08. Archived from the original (http://www.rnhit.com/mcns/articles/commander-in-chief.html?PHPSESSID=8a9729bc19d560d81319bc8bee3f380a) on 2007-09-28. . Retrieved 2007-04-20.

[8] http://www.colombopage.com/archive_10A/May15_1273933283JR.php

[9] "Sri Lanka Army History, The Ancient Time" (http://web.archive.org/web/20070202064404/http://www.army.lk/history2.php). *Sri Lanka Army*. Archived from the original (http://www.army.lk/history2.php) on 2007-02-02. . Retrieved 2007-02-04.

[10] Army, Sri Lanka (October 1999). "Chapter 1". *Sri Lanka Army,* "50 YEARS ON" - 1949-1999 (1st Edition ed.). Colombo: Sri Lanka Army. pp. 1–3. ISBN 978-955-8089-02-6.

[11] "The Portuguese in Sri Lanka (1505-1658)" (http://www.lankalibrary.com/geo/portu2.htm). *WWW Virtual Library - Sri Lanka*. . Retrieved 2004-06-30.

[12] "Indias.com, Sri Lankan Army" (http://www.indias.com/php/wiki-Sri_Lankan_Army). *Indias*. . Retrieved 2006-02-04.

[13] "History of the Sri Lanka Light Infantry Regiment, Sri Lanka Army" (http://web.archive.org/web/20060708103051/http://www.army.lk/lightinfan.php). *Sri Lanka Army*. Archived from the original (http://www.army.lk/lightinfan.php) on 2006-07-08. . Retrieved 2006-02-04.

[14] "Establishment, Sri Lanka Army" (http://web.archive.org/web/20060326024640/http://www.army.lk/org.php). *Sri Lanka Army*. Archived from the original (http://army.lk/org.php) on 2006-03-26. . Retrieved 2006-02-04.

[15] Sergei de Silva-Ranasinghe looks back at the early days of the Sri Lanka Army (http://www.sundaytimes.lk/051009/plus/1.html)

[16] An evolving army and its role through time, Sergei de Silva- Ranasinghe's article on the early days of the Sri Lanka Army (http://www.sundaytimes.lk/051016/plus/4.html)

[17] "Sri Lanka Army Marks 50 Years" (http://www.washingtonpost.com/wp-srv/aponline/19991010/aponline085414_000.htm). Washingtonpost, AP News. October 10, 1999. .

[18] Sri Lanka's Tamil Tigers admit defeat (http://www.csmonitor.com/2009/0517/p90s01-wosc.html) Christian Science Monitor - May 17, 2009

[19] "Sri Lanka Light Infantry UN Peacekeeping Operations" (http://web.archive.org/web/20070204125105/http://www.army.lk/lightinfan.php). *Sri Lanka Army*. Archived from the original (http://www.army.lk/lightinfan.php) on 2007-02-04. . Retrieved 2007-03-16.

[20] "2 U.N. peacekeepers killed in Haiti fighting" (http://www.signonsandiego.com/uniontrib/20050321/news_1n21world.html). San Diego Union Tribune. March 21, 2005. .

[21] Sri Lankan troops in Haiti honoured (http://www.dailynews.lk/2008/04/02/news13.asp)

[22] "Haiti: Over 100 Sri Lankan blue helmets repatriated on disciplinary grounds – UN" (http://www.un.org/apps/news/ticker/tickerstory.asp?NewsID=24514). *United Nations*. 2007-11-02. . Retrieved 2007-11-04.

[23] Williams, Carol J. (2007-12-15). "U.N. confronts another sex scandal" (http://articles.latimes.com/2007/dec/15/world/fg-haitisex15). *Los Angeles Times*. . Retrieved 2008-11-07.

[24] "Sri Lanka to probe UN sex claims" (http://news.bbc.co.uk/2/hi/south_asia/7076284.stm). *BBC*. 2007-11-03. . Retrieved 2007-11-05.

[25] Reddy, B. Muralidhar (2007-11-05). "Part of Sri Lankan contingent in Haiti to be sent back" (http://www.hindu.com/2007/11/04/stories/2007110457201800.htm). *The Hindu*. . Retrieved 2007-11-05.

[26] "UN confirms sex charges" (http://www.sundaytimes.lk/080330/News/news0015.html). *Sundaytimes*. 2008-03-30. . Retrieved 2008-03-30.

[27] Haiti - MINUSTAH - Facts and Figures (http://www.un.org/depts/dpko/missions/minustah/facts.html)

[28] http://www.army.lk/detailed.php?NewsId=2195

[29] http://www.dailymirror.lk/print/index.php/news/front-page-news/11435-army-contingent-to-chad-under-un-flag.html

[30] Lankan troops for Lebanon UN peace keeping mission (http://www.island.lk/index.php?page_cat=article-details&page=article-details&code_title=11739)

[31] "Military Balance, Sri Lanka" (http://web.archive.org/web/20061002155351/http://www.ipcs.org/Military_militaryBalance2.jsp?database=1002&country2=Sri+Lanka). *Institute of Peace and Conflict Studies*. Archived from the original (http://www.ipcs.org/Military_militaryBalance2.jsp?database=1002&country2=Sri Lanka) on 2006-10-02. . Retrieved 2007-03-14.

[32] "Sri Lanka, The Army" (http://www.photius.com/countries/sri_lanka/national_security/sri_lanka_national_security_the_army.html). *Photius Coutsoukis*. . Retrieved 2007-03-14.

[33] Two Security Forces Headquarters established in Mullaitivu, Kilinochchi (http://www.dailynews.lk/2009/06/01/sec02.asp)

[34] Security Forces on a realistic path to achieve the set target, Dailynews (http://www.dailynews.lk/2008/07/14/spo10.asp)

[35] Bitter battles, heavy toll, Tiger strongholds under siege (http://www.sundaytimes.lk/081123/Columns/sitreport.html)

[36] Two New SF Headquarters Become Operational (http://www.army.lk/detailed.php?NewsId=697)

[37] New name for JOH (http://www.sundaytimes.lk/090927/News/news_22.html)

[38] Sixth Security Forces Headquarters to Take Charge of Security in South (http://www.army.lk/detailed.php?NewsId=1287)

[39] Eelam War IV: Tigers Face Defeat (http://www.army.lk/morenewsfet.php?id=5272)

[40] "Training, Sri Lanka Army" (http://web.archive.org/web/20060410062148/http://www.army.lk/train.php). *Sri Lanka Army*. Archived from the original (http://army.lk/train.php) on 2006-04-10. . Retrieved 2006-02-04.

[41] "Sri Lanka" (http://www.nationsencyclopedia.com/Asia-and-Oceania/Sri-Lanka-ARMED-FORCES.html). *Nations Encyclopedia*. . Retrieved 2007-02-04.

[42] 'Tamil unit' for Sri Lanka's army (http://news.bbc.co.uk/2/hi/south_asia/8124836.stm) BBC - June 29, 2009

[43] "Victory's price: 6,200 Sri Lankan troops" (http://news.smh.com.au/breaking-news-world/victorys-price-6200-sri-lankan-troops-20090522-bi4f.html). News.smh.com.au. 22 May 2009. . Retrieved 30 May 2009.

[44] Crushing the revolt (http://sundaytimes.lk/010408/spec.html)

[45] Kanagaraarachchi, Ramani (2007-01-20). "Heroes who made the supreme sacrifice" (http://www.dailynews.lk/2007/01/20/spice01.asp). *Daily News* (Associated Newspapers of Ceylon). . Retrieved 2007-05-25.

[46] Heroes who made the supreme sacrifice (http://www.dailynews.lk/2007/01/20/spice01.asp)

[47] Ratnayake, Brigadier Daya (2005-07-30). "Noble in death as in life" (http://sundaytimes.lk/050703/plus/appreciation.html). *The Sunday Times* (Wijeya Newspapers Ltd). . Retrieved 2007-08-18.

[48] De Silva, Senaka (2005-06-01). "Senior Army officer shot dead" (http://web.archive.org/web/20070930224634/http://www.dailymirror.lk/2005/06/01/front/1.asp). *Daily Mirror* (Wijeya Newspapers Ltd). Archived from the original (http://www.dailymirror.lk/2005/06/01/front/1.asp) on 2007-09-30. . Retrieved 2007-06-03.

[49] "Sri Lankan Army Women's Corps" (http://womenshistory.about.com/library/ency/blwh_sri_lanka_women_military.htm). *About, Inc.* . Retrieved 2007-02-04.

[50] "An officer and a lady; You've come a long way, lass." (http://www.sundayobserver.lk/2006/10/08/zin01.asp). *Sunday Observer*. . Retrieved 2006-10-08.

[51] "Sri Lanka Army Women's Corps, Sri Lanka Army" (http://web.archive.org/web/20070128135530/http://www.army.lk/regiment_regular_19.php). *Sri Lanka Army*. Archived from the original (http://www.army.lk/regiment_regular_19.php) on 2007-01-28. . Retrieved 2007-02-04.

[52] Michael K. Cecil - Sri Lanka's Military: The Search For A Mission, (http://www.bdblodgett.com/srilanka.pdf).

[53] Army, Sri Lanka (October 1999). "Chapter 1". *Sri Lanka Army,* "50 YEARS ON" - 1949-1999 (1st Edition ed.). Colombo: Sri Lanka Army. p. 393. ISBN 978-955-8089-02-6.

[54] http://www.army.lk/org2.php

[55] "SFrom Unicorn to Unibuffel" (http://www.sundayobserver.lk/2004/04/25/fea22.html). sundayobserver. .

[56] Bedi, Rahul (2007-06-02). "Sri Lanka turns to Pakistan, China for military needs" (http://www.indianmuslims.info/news/2007/jun/02/sri_lanka_turns_pakistan_china_military_needs.html). *IANS* (Urdustan.com Network). . Retrieved 2007-06-02.

[57] "Small Arms and Light Weapons Conference Report" (http://www.globalactionpw.org/un/SALW_PoA_RC_ REPORT.pdf) (PDF). *Judith McDaniel*. . Retrieved 2006-07-23.

[58] "No Nonsense guide to The Arms Trade" (http://www.thirdworldtraveler.com/Weapons/Arms_Trade.html). *Gideon Burrows*. . Retrieved 2005-11-19.

[59] "Naval surveillance is the millstone around LTTE's neck" (http://www.priu.gov.lk/news_update/features/20031017naval_surveillance_millstone_LTTE.htm). *P.K. Balachanddran*. . Retrieved 2003-10-17.

[60] "Type 69 Main Battle Tank" (http://www.globalsecurity.org/military/world/china/type-69.htm). GlobalSecurity.org. July 31, 2005. . Retrieved 2007-03-15.

[61] "Sri Lanka (LTTE)" (http://acd.iiss.org/armedconflict/Mainpages/dsp_ConflictWeapons.asp?ConflictID=174). International Institute For Strategic Studies. March 15, 2007. . Retrieved 2007-03-15.

[62] "ZSD89 Armoured Personnel Carrier" (http://www.sinodefence.com/army/armour/zsd89.asp). sinodefence.com. 2007. . Retrieved 2007-12-02.

[63] [//www.nation.lk/2006/10/15/newsfe1.htm "Type 63"]. nation.lk. //www.nation.lk/2006/10/15/newsfe1.htm.

[64] "United Nations Register of Conventional Arms" (http://disarmament.un.org/library.nsf/c793d171848bac2b85256d7500700384/c7b9a51a4973f6e585256dc100585789/$FILE/sg58.203.pdf) (PDF). *United Nations*. August 21, 2003. . Retrieved 2007-03-16.

[65] Saferworld's research project on arms and security in EU Associate Countries, Czech Republic (http://www.saferworld.org.uk/images/pubdocs/Beast CzechRep.pdf)

[66] Smith, Chris (October 2003) (PDF). *In the Shadow of a Cease-fire: The Impacts of Small Arms Availability and Misuse in Sri Lanka* (http://www.smallarmssurvey.org/files/sas/publications/o_papers_pdf/2003-op11-sri_lanka.pdf). Small Arms Survey. .

[67] "T 81 Assault Rifle" (http://web.archive.org/web/20070214094032/http://www.sinodefence.com/army/small_arms/type81rifle.asp). *SinoDefence*. Archived from the original (http://www.sinodefence.com/army/small_arms/type81rifle.asp) on 2007-02-14. . Retrieved 2007-03-14.

[68] International, Amnesty; International Secretariat, AI Ireland, Omega Foundation (PDF). *Undermining Global Security: the European Union's arms exports* (http://www.smallarmssurvey.org/files/portal/spotlight/ammunition/ammun_pdf/2004_amnesty_international.pdf) (1st Edition ed.). Amnesty International. ISBN 0-86210-356-8. . Retrieved 14 March 2007.

[69] "Profiling the Small Arms Industry" (http://www.worldpolicy.org/projects/arms/reports/smallarms.htm). *Arms Trade Resource Center*. World Policy Institute. November 2000. . Retrieved 2007-03-15.

[70] "Accuracy International L96A1" (http://www.militaryfactory.com/smallarms/detail.asp?smallarms_id=34). www.MilitaryFactory.com. January 16, 2009. . Retrieved 2009-01-20.

[71] "Trans-National Crime and Light Weapons Proliferation: Security Implications for the State" (http://web.archive.org/web/20061207164518/http://www.cpdsindia.org/globalhumansecurity/trans-national.htm). *Tara Kartha*. IDSA. August 15, 2002. Archived from the original (http://www.cpdsindia.org/globalhumansecurity/trans-national.htm) on 2006-12-07. . Retrieved 2007-03-15.

[72] (http://www.sipri.org/research/armaments/transfers/transparency/national_reports/serbia/SER_08.pdf/view)

[73] "Biography of Mikhail Kalashnikov" (http://www.vodkakalashnikov.com/biography.htm). VRQ International, Inc.. January 14, 2004. . Retrieved 2007-03-15.

[74] "If You're Involved in Resolving the" (http://www.ibiblio.org/obl/reg.burma/archives/199901/msg00057.html). ibiblio.org. January 5, 1999. . Retrieved 2007-03-15.

72. http://www.spur.asn.au/extra/Curr10.htm

External links

- Official website (http://www.army.lk)
- Ministry of Defence Sri Lanka (http://www.defence.lk/)
- General Sir John Kotelawala Defence Academy (http://www.kda.lk/)

Article Sources and Contributors

Sri_Lanka_Artillery *Source*: http://en.wikipedia.org/w/index.php?title=Sri_Lanka_Artillery *Contributors*: Arch dude, Buistr, Chamal N, Chanakal, CommonsDelinker, Cossde, Cremepuff222, Dakz007, Gayal, Geniac, Gira2be, Jaraalbe, Lahiru k, M-le-mot-dit, Netmonger, Nitraven, Quadell, R'n'B, Rich Farmbrough, Rjwilmsi, Roleplayer, Sphilbrick, Sujeewa7770, 13 anonymous edits

Sri_Lanka_Army *Source*: http://en.wikipedia.org/w/index.php?title=Sri_Lanka_Army *Contributors*: 007Explorer, Alarics, Aldis90, Alison, Askari Mark, Astronomyinertia, BarretB, Barticus88, Black Falcon, Blackknight12, Bobblehead, Bsadowski1, Butseriouslyfolks, CBM, CSWarren, CWesling, Caniago, Chamal N, Chanakal, ChemGardener, Chidis, Chris the speller, CommonsDelinker, Cossde, DHeyward, DMI, Deathbunny, Debresser, Defenderline, Derekbridges, Distributor108, Dnv-kindredblades, Dormskirk, Drbreznjev, ESkog, Ebaran, Editore99, Eelampulikal, El Zoof, Elalan, Errabee, Euchiasmus, Feroshki, Firewater101, Flynx22, Friism, Fritz Saalfeld, Fuad masrur, Geniac, GiW, Gira2be, Giraffedata, Gogo Dodo, GoingBatty, Ground Zero, Halcionne, HarryHenryGebel, Haus, Hibernian, Hmains, Hmwith, Hu12, Icairns, Icemansatriani, Imaninjapirate, Iwazaki, J.smith, Ja 62, Japuraalwis, Jaraalbe, Jeff3000, Jesse Viviano, John of Reading, Joseph1999, Jpoelma13, Kanatonian, Kerr avon, Laaa200, Lahiru k, Lakdiva, Lanka07, Lankaupdate, Lexicon, LilHelpa, M3mad, Manjula.aw, Marinecore88, Maxim, Melesse, Memilanka, Milton Stanley, Miyagawa, MoRsE, NavodEranda, Netmonger, Neuralolive, Nick Number, NinjaLOD, Nitraven, Obi2canibe, Ominae, Open2universe, OverlordQ, PachchaSira, Pectore, Peripitus, Pharaoh of the Wizards, Phil Bridger, Puddhe, Pulasti, Qxz, R'n'B, RG2, Rajithcalwis, Raptor ams, Rbraunwa, Rettetast, Rich Farmbrough, Rjwilmsi, RoyBoy, Rumpelstiltskin223, SBR249, Sankili kumar, Sarvagnya, SebastianHelm, Sechzehn, Segregator236, Shan246, Shaxdev12, Shaymtnwalker, Shlomke, Shoessss, Sinhala freedom, Skumarla, SlackerMom, Smgunasekara, Snowolfd4, Spinach Monster, Steven J. Anderson, Sudharsansn, THEN WHO WAS PHONE?, Thamira80, The Thing That Should Not Be, Thuan.salim, Tmaull, Tyronen, Ulric1313, Umapathy, Vanished 6551232, Versus22, Vipinhari, Wally, Watchdogb, Welsh, Who then was a gentleman?, WikHead, Wiki Raja, Woohookitty, Yucatann, Zonda2323, 225 anonymous edits

Panagoda_Cantonment *Source*: http://en.wikipedia.org/w/index.php?title=Panagoda_Cantonment *Contributors*: Chanakal, Cossde, Gira2be, Jevansen, Lahiru k, Nitraven, Obi2canibe, TopoChecker, 1 anonymous edits

Artillery *Source*: http://en.wikipedia.org/w/index.php?title=Artillery *Contributors*: 1 + 2 x 8, 62.253.64.xxx, AD66, Abnoas, Abqsteve, Acroterion, Addboy, Aeonx, Airborne84, Alan Parmenter, Aldis90, AlexiusHoratius, Alexthe5th, Alheidg, AllStarZ, Altzinn, Alureiter, Andonic, AndreyA, AngBent, ArglebargleIV, Art LaPella, Artene50, Artillerie, Askari Mark, Attilios, Axeman89, Az81964444, B.Soto, BD2412, Bazonka, Beetstra, Bejnar, Beland, Betacommand, Beyond My Ken, Bjenks, Blackshod, BlaiseFEgan, Bleakcomb, Bloodshedder, Bluemoose, Bob the Cannibal, Bobblewik, Bogsat, Boulevardier, Brandy Frisky, Brim, Bronco, Bryan Derksen, Bryan Nguyen, Bryson109, Buchraeumer, Buckshot06, Bullzeye, Bumble bee, Bumm13, Buonaparte69, Byrnejb, CIreland, CWenger, Cabalamat, Calzarette, CambridgeBayWeather, Cancun771, CardinalDan, Carl.bunderson, Camildo, Cfrydj, Chairboy, Chessy999, ChewyLSB, Chris5369, Cinnamon42, Cjkporter, Cjthellama, ClemsonTiger, Cliché Online, Closedmouth, Coga2006, Colin Douglas Howell, Color probe, CommonsDelinker, Conscious, ContiAWD, Conversion script, CoolFox, Coolhawks88, Coughinink, CrankyScorpion, Ctmt, Cyclopaedic, Cyde, Cyrius, D6, DITWIN GRIM, DMorpheus, DVD R W, Daemonofdecay, DanMS, DancingPenguin, Danielx, Darklilac, Darth Marik87, David.Monniaux, Dbachmann, Descendall, Dialectric, DocKrin, DocWatson42, Docu, DougsTech, Dr. VanNostrand, DragonHawk, Dsmdgold, Dungodung, Dylan Hsu, Dystopos, EHFOH, Earendur, Electron9, Eleland, Ellsworth, EmanWilm, Enarei, Enchanter, Enriquecornejo, EoGuy, Ericvigil, Erudy, Erzbischof, Europrobe, Everyking, Facepalming, Falcon8765, Fancy steve, Felis domestica, Finn-Zoltan, Firepower Canada, Fix Bayonets!, Floridan, Fountains of Bryn Mawr, Fredrik, G27, GCarty, GMan552, Gaius Cornelius, Gene Nygaard, Gh5046, Ghewgill, Gire 3pich2005, Glenn, Gob Lofa, GoneAwayNowAndRetired, GraemeLeggett, Grafen, Grampion76, GreatWhiteNortherner, Gregorydavid, Groundsquirrel13, Gsl, Guapovia, Gun Powder Ma, Hadal, Hairy Dude, Herbertxu, Hibernian, HighKing, Hirsch.im.wald, Hmains, Hohum, Howcheng, Htra0497, Hutcher, Hux, IVAN3MAN, Ian Pitchford, Ianjohnstewart, Iceberg3k, Immunize, Indefatigable, Ingolfson, Instinct, Iohannes Animosus, Iridescent, J-stan, J.delanoy, Jagged 85, JamesAM, Jan van Male, Janus Shadowsong, Jbowman90, Jcw69, Jeepday, Jeff F F, Jezarnold, Jfdwolff, Jhutzler, Jigen III, Jmmalek, JoeSmack, Johan1298, Jolomo, Jose77, Josephs1, Joshbaumgartner, Jpbrenna, Jrleighton, Jrockley, Julesd, KTo288, Kathryn NicDhàna, Keegan, KelleyCook, Ken.guo, Kenneth Smelley, Ketiltrout, Kevin, Kevin Saff, Kevinusmartinus, Klemen Kocjancic, Koakhtzvigad, Kormoran, Krisbaum, Kross, Ksyrie, Kubanczyk, Kumioko, Kwamikagami, L0b0t, L3th4l, LWF, Legaiaflame, Leibniz, Leutha, LiDaobing, Lightmouse, LittleOldMe, Looper5920, Lotje, LtNOWIS, MCG, MER-C, Mad-mckechnie, Maelnuneb, Magister Mathematicae, Magnus Manske, Manway, Manxruler, Marcus Qwertyus, Martynas Patasius, MathKnight, Mathiasrex, MatthewVanitas, Maurreen, Max (the terminator), Mazca, Mentifisto, Miaow Miaow, Michael Hardy, Michael Zimmermann, Mikeo, Milesli, Miskatonic, Mkweise, MoRsE, Montgomery '39, Motor.on, Motorfix, Mozkill, Mrg3105, Mtnerd, Mumia-w-18, NSK Nikolaos S. Karastathis, Nabokov, Nagy, Nantuk, NarayanGa, NawlinWiki, Nfe, Nk, Nobunaga24, Nohomers48, NorsemanII, Nvinen, Oden, Ohnoitsjamie, Olegwiki, Onthegogo, Optimist on the run, Orpheus, Palm dogg, Parsecboy, Patrick, Paulnusbaum, Pekkapihlajasaari, PeruvianIlama, Peter Chamberlain, Peter Isotalo, Petri Krohn, Phatom87, PhilKnight, Philip Trueman, Pibwl, Piotrus, Pol098, Portalian, Poter99, Primalchaos, Purplepox01, QuentinV, Qui1che, RJASE1, Rama, Raul654, Rebutcher, RedStarFire, Reedy, Rich Farmbrough, RickScott, Riddley, Rjwilmsi, Rl, Rmoyes, Rock2e, Roger Davies, Rossami, Rrburke, Rusty2005, S h i v a (Visnu), Salmanazar, Sc147, Sean.nobles, Securiger, Sekigaara, Sgm2000, Shadowjams, Shark kid, Sherurcij, Shlomke, SimonATL, Sindhbadh, Skiracer4, SkyWalker, Snapperman2, SnowFire, Soap, Socrates2008, Sophysduckling, SpookyMulder, SunCreator, SwordSmurf, Symane, TenPoundHammer, The Land, The Little Internet Kitty, The Nut, The Thing That Should Not Be, The one that edits, Therealhazel, Theshocker3, Tiniti, Tobby72, Tom harrison, TomStar81, Tpbradbury, Tracy49, Trainik, Trekphiler, Trovatore, TwoOneTwo, TykkimiesPSPR, Ulterior19802005, UltimateDestroyerOfWorlds, Uniquely Fabricated, Urhixidur, Uriah923, Varlet16, Vietbio, Vriullop, Wa3frp, Wayward, Wendell, Wikiliki, Wikist, Wimt, Woohookitty, X96lee15, Xezbeth, Xtzou, Yakamoz51, Yurik, Zahid Abdassabur, Zanoni, Zephyrus67, ZeroOne, Zigger, Zilyuki, Zonglowe, 415 anonymous edits

Sri_Lanka_Army_Order_of_Precedence *Source*: http://en.wikipedia.org/w/index.php?title=Sri_Lanka_Army_Order_of_Precedence *Contributors*: Gira2be, Lahiru k, LtNOWIS, Nitraven

Sri_Lanka_Armed_Forces *Source*: http://en.wikipedia.org/w/index.php?title=Sri_Lanka_Armed_Forces *Contributors*: Afinebalance, Alansohn, AnthonyA7, AuburnPilot, BWD, BarretB, Bastin, Benlisquare, Benvogel, Big Bob the Finder, Black Falcon, Blackknight12, Bogdan, CYMikeLee, Caerwine, Calliopejen1, Chamal N, Chidis, Chontoman, Closedmouth, Cohesion, CommonsDelinker, Conversion script, CooldogCongo, Cossde, DARTH SIDIOUS 2, DarkFalls, David Newton, Defenderline, Diannaa, Distributor108, Dr.K., Dylan620, Elalan, Estrellador*, Geniac, Gira2be, Grafen, Graham87, Greenshed, Hadooookin, Hazard-SJ, Hdt83, Hitmanrock, Hmains, Hojimachong, HudsonBreeze, Icemansatriani, Idleguy, Intoronto1125, J.delanoy, Joel7687, Justin Herbert, Karl Dickman, Lahiru k, Lankaupdate, Levekks, Lexicon, Lkinkade, Lokal Profil, Lronhubbard234, LtNOWIS, Mdd4696, Melee Clark, Memilanka, MentalDimension, Nitraven, Obi2canibe, PBP, Palaly, Pathoschild, PhnomPencil, Pulasti, RL0919, Redvers, Reenem, Rehman, Rettetast, Richard001, Ricky81682, Rjwilmsi, Rlandmann, SJay 2007, SachiniFernando, Saypotatoes3, Seano1, SebastianHelm, Senthilwikiacc, Sentinel R, Shan246, Simon12, Sir Nicholas de Mimsy-Porpington, Snowolfd4, Superbeecat, Syrthiss, Tamiltony, Theredstarswl, UN1, Ukumaran10, Vanished 6551232, Watchdogb, WereSpielChequers, Wiki Raja, Woohookitty, YellowMonkey, Yoshi:3, Yucatann, 120 anonymous edits

List_of_Sri_Lanka_Army_regiments_and_corps *Source*: http://en.wikipedia.org/w/index.php?title=List_of_Sri_Lanka_Army_regiments_and_corps *Contributors*: Buckshot06, Chamal N, Gira2be, Mrg3105, Nitraven, PuppyGirl1508, Seba5618, 1 anonymous edits

Sri Lanka Army *Source*: http://en.wikipedia.org/w/index.php?title=Sri_Lanka_Army *Contributors*: 007Explorer, Alarics, Aldis90, Alison, Askari Mark, Astronomyinertia, BarretB, Barticus88, Black Falcon, Blackknight12, Bobblehead, Bsadowski1, Butseriouslyfolks, CBM, CSWarren, CWesling, Caniago, Chamal N, Chanakal, ChemGardener, Chidis, Chris the speller, CommonsDelinker, Cossde, DHeyward, DMI, Deathbunny, Debresser, Defenderline, Derekbridges, Distributor108, Dnv-kindredblades, Dormskirk, Drbreznjev, ESkog, Ebaran, Editore99, Eelampulikal, El Zoof, Elalan, Errabee, Euchiasmus, Feroshki, Firewater101, Flynx22, Friism, Fritz Saalfeld, Fuad masrur, Geniac, GiW, Gira2be, Giraffedata, Gogo Dodo, GoingBatty, Ground Zero, Halcionne, HarryHenryGebel, Haus, Hibernian, Hmains, Hmwith, Hu12, Icairns, Icemansatriani, Imaninjapirate, Iwazaki, J.smith, Ja 62, Japuraalwis, Jaraalbe, Jeff3000, Jesse Viviano, John of Reading, Joseph1999, Jpoelma13, Kanatonian, Kerr avon, Laaa200, Lahiru k, Lakdiva, Lanka07, Lankaupdate, Lexicon, LilHelpa, M3mad, Manjula.aw, Marinecore88, Maxim, Melesse, Memilanka, Milton Stanley, Miyagawa, MoRsE, NavodEranda, Netmonger, Neuralolive, Nick Number, NinjaLOD, Nitraven, Obi2canibe, Ominae, Open2universe, OverlordQ, PachchaSira, Pectore, Peripitus, Pharaoh of the Wizards, Phil Bridger, Puddhe, Pulasti, Qxz, R'n'B, RG2, Rajithcalwis, Raptor ams, Rbraunwa, Rettetast, Rich Farmbrough, Rjwilmsi, RoyBoy, Rumpelstiltskin223, SBR249, Sankili kumar, Sarvagnya, SebastianHelm, Sechzehn, Segregator236, Shan246, Shaxdev12, Shaymtnwalker, Shlomke, Shoessss, Sinhala freedom, Skumarla, SlackerMom, Smgunasekara, Snowolfd4, Spinach Monster, Steven J. Anderson, Sudharsansn, THEN WHO WAS PHONE?, Thamira80, The Thing That Should Not Be, Thuan.salim, Tmaull, Tyronen, Ulric1313, Umapathy, Vanished 6551232, Versus22, Vipinhari, Wally, Watchdogb, Welsh, Who then was a gentleman?, WikHead, Wiki Raja, Woohookitty, Yucatann, Zonda2323, 225 anonymous edits

Image Sources, Licenses and Contributors

File:SLA Artillery RM70 rocket launcher.JPG *Source*: http://en.wikipedia.org/w/index.php?title=File:SLA_Artillery_RM70_rocket_launcher.JPG *License*: unknown *Contributors*: User:Chamal N

File:Flag of the United Kingdom.svg *Source*: http://en.wikipedia.org/w/index.php?title=File:Flag_of_the_United_Kingdom.svg *License*: unknown *Contributors*: Anomie, Mifter

File:Flag of India.svg *Source*: http://en.wikipedia.org/w/index.php?title=File:Flag_of_India.svg *License*: unknown *Contributors*: Anomie, Mifter

Image:Sri Lanka Army Logo.png *Source*: http://en.wikipedia.org/w/index.php?title=File:Sri_Lanka_Army_Logo.png *License*: unknown *Contributors*: Geniac, Lahiru k, Nitraven

Image:The Sri Lanka Army Flag And Crest.JPG *Source*: http://en.wikipedia.org/w/index.php?title=File:The_Sri_Lanka_Army_Flag_And_Crest.JPG *License*: unknown *Contributors*: Cossde, Geniac, Goustien, Lahiru k, Nitraven

Image:Flag of Sri Lanka.svg *Source*: http://en.wikipedia.org/w/index.php?title=File:Flag_of_Sri_Lanka.svg *License*: unknown *Contributors*: Zscout370

File:PM visits the 1st Bn CLI.jpg *Source*: http://en.wikipedia.org/w/index.php?title=File:PM_visits_the_1st_Bn_CLI.jpg *License*: unknown *Contributors*: Magog the Ogre, Wolkodlak

Image:SLA-50.jpg *Source*: http://en.wikipedia.org/w/index.php?title=File:SLA-50.jpg *License*: unknown *Contributors*: Martin H., Wolkodlak

Image:SLAC-1-.png *Source*: http://en.wikipedia.org/w/index.php?title=File:SLAC-1-.png *License*: unknown *Contributors*: User:Lahiru_k

Image:SLE.png *Source*: http://en.wikipedia.org/w/index.php?title=File:SLE.png *License*: unknown *Contributors*: User:Lahiru_k

Image:SLSC.png *Source*: http://en.wikipedia.org/w/index.php?title=File:SLSC.png *License*: unknown *Contributors*: User:Lahiru_k

Image:SLLI Crest.gif *Source*: http://en.wikipedia.org/w/index.php?title=File:SLLI_Crest.gif *License*: unknown *Contributors*: Cmdrjameson, Geniac, Nitraven

Image:SLSR.png *Source*: http://en.wikipedia.org/w/index.php?title=File:SLSR.png *License*: unknown *Contributors*: User:Lahiru_k

Image:GW.png *Source*: http://en.wikipedia.org/w/index.php?title=File:GW.png *License*: unknown *Contributors*: User:Lahiru_k

Image:GR.png *Source*: http://en.wikipedia.org/w/index.php?title=File:GR.png *License*: unknown *Contributors*: User:Lahiru_k

Image:VIR.png *Source*: http://en.wikipedia.org/w/index.php?title=File:VIR.png *License*: unknown *Contributors*: User:Lahiru_k

Image:SLASC.png *Source*: http://en.wikipedia.org/w/index.php?title=File:SLASC.png *License*: unknown *Contributors*: User:Lahiru_k

Image:SLAMC.png *Source*: http://en.wikipedia.org/w/index.php?title=File:SLAMC.png *License*: unknown *Contributors*: User:Lahiru_k

Image:SLAOC.png *Source*: http://en.wikipedia.org/w/index.php?title=File:SLAOC.png *License*: unknown *Contributors*: User:Lahiru_k

Image:SLCMP.png *Source*: http://en.wikipedia.org/w/index.php?title=File:SLCMP.png *License*: unknown *Contributors*: User:Lahiru_k

Image:SLAGSC.png *Source*: http://en.wikipedia.org/w/index.php?title=File:SLAGSC.png *License*: unknown *Contributors*: User:Lahiru_k

Image:SLARC.png *Source*: http://en.wikipedia.org/w/index.php?title=File:SLARC.png *License*: unknown *Contributors*: User:Lahiru_k

Image:SLAPC.png *Source*: http://en.wikipedia.org/w/index.php?title=File:SLAPC.png *License*: unknown *Contributors*: User:Lahiru_k

File:Flag of the Soviet Union.svg *Source*: http://en.wikipedia.org/w/index.php?title=File:Flag_of_the_Soviet_Union.svg *License*: unknown *Contributors*: A1, Ahmadi, Alex Smotrov, Alvis Jean, Art-top, BagnoHax, Brandmeister, Denniss, ELeschev, Endless-tripper, EugeneZelenko, F l a n k e r, Fred J, Fry1989, G.dallorto, Garynysmon, Herbythyme, Homo lupus, Jake Wartenberg, MaggotMaster, Ms2ger, Nightstallion, Pianist, R-41, Rainforest tropicana, Sebyugez, Solbris, Storkk, Str4nd, Tabasco, ThomasPusch, Toben, Twilight Chill, Xgeorg, Zscout370, Серп, Тоньч4, 59 anonymous edits

File:Flag of the People's Republic of China.svg *Source*: http://en.wikipedia.org/w/index.php?title=File:Flag_of_the_People's_Republic_of_China.svg *License*: unknown *Contributors*: User:Denelson83, User:SKopp, User:Shizhao, User:Zscout370

File:Flag of Sri Lanka.svg *Source*: http://en.wikipedia.org/w/index.php?title=File:Flag_of_Sri_Lanka.svg *License*: unknown *Contributors*: Zscout370

File:Flag of South Africa.svg *Source*: http://en.wikipedia.org/w/index.php?title=File:Flag_of_South_Africa.svg *License*: unknown *Contributors*: Adriaan, Anime Addict AA, AnonMoos, BRUTE, Daemonic Kangaroo, Dnik, Duduziq, Dzordzm, Fry1989, Homo lupus, Jappalang, Juliancolton, Kam Solusar, Klemen Kocjancic, Klymene, Lexxyy, Mahahahaneapneap, Manuelt15, Moviedefender, NeverDoING, Ninane, Poznaniak, Przemub, SKopp, ThePCKid, ThomasPusch, Tvdm, Ultratomio, Vzb83, Zscout370, 34 anonymous edits

File:Flag of Czechoslovakia.svg *Source*: http://en.wikipedia.org/w/index.php?title=File:Flag_of_Czechoslovakia.svg *License*: unknown *Contributors*: (of code)

File:Flag of the United States.svg *Source*: http://en.wikipedia.org/w/index.php?title=File:Flag_of_the_United_States.svg *License*: unknown *Contributors*: Anomie

File:Flag of Italy.svg *Source*: http://en.wikipedia.org/w/index.php?title=File:Flag_of_Italy.svg *License*: unknown *Contributors*: Anomie

File:Flag of Austria.svg *Source*: http://en.wikipedia.org/w/index.php?title=File:Flag_of_Austria.svg *License*: unknown *Contributors*: User:SKopp

File:Flag of Germany.svg *Source*: http://en.wikipedia.org/w/index.php?title=File:Flag_of_Germany.svg *License*: unknown *Contributors*: Anomie

File:Flag of Belgium (civil).svg *Source*: http://en.wikipedia.org/w/index.php?title=File:Flag_of_Belgium_(civil).svg *License*: unknown *Contributors*: Bean49, David Descamps, Dbenbenn, Denelson83, Evanc0912, Fry1989, Gabriel trzy, Howcome, Ms2ger, Nightstallion, Oreo Priest, Rocket000, Rodejong, Sir Iain, ThomasPusch, Warddr, Zscout370, 4 anonymous edits

File:Flag of Singapore.svg *Source*: http://en.wikipedia.org/w/index.php?title=File:Flag_of_Singapore.svg *License*: unknown *Contributors*: Various

File:Flag of Israel.svg *Source*: http://en.wikipedia.org/w/index.php?title=File:Flag_of_Israel.svg *License*: unknown *Contributors*: AnonMoos, Bastique, Bobika, Brown spite, Captain Zizi, Cerveaugenie, Drork, Etams, Fred J, Fry1989, Geagea, Himasaram, Homo lupus, Humus sapiens, Klemen Kocjancic, Kookaburra, Luispihormiguero, Madden, Neq00, NielsF, Nightstallion, Oren neu dag, Patstuart, PeeJay2K3, Pumbaa80, Ramiy, Reisio, Rodejong, SKopp, Sceptic, SomeDudeWithAUserName, Technion, Typhix, Valentinian, Yellow up, Zscout370, 31 anonymous edits

File:Flag of Serbia.svg *Source*: http://en.wikipedia.org/w/index.php?title=File:Flag_of_Serbia.svg *License*: unknown *Contributors*: sodipodi.com

File:Flag of Russia.svg *Source*: http://en.wikipedia.org/w/index.php?title=File:Flag_of_Russia.svg *License*: unknown *Contributors*: Anomie

File:Flag of Pakistan.svg *Source*: http://en.wikipedia.org/w/index.php?title=File:Flag_of_Pakistan.svg *License*: unknown *Contributors*: Zscout370

File:MuseeMarine-canon-1880-p1000439.jpg *Source*: http://en.wikipedia.org/w/index.php?title=File:MuseeMarine-canon-1880-p1000439.jpg *License*: unknown *Contributors*: User:Rama

File:French soldiers in the Franco-Prussian War 1870-71.jpg *Source*: http://en.wikipedia.org/w/index.php?title=File:French_soldiers_in_the_Franco-Prussian_War_1870-71.jpg *License*: unknown *Contributors*: DIREKTOR, Ribberlin

Image:M777 Light Towed Howitzer 1.jpg *Source*: http://en.wikipedia.org/w/index.php?title=File:M777_Light_Towed_Howitzer_1.jpg *License*: unknown *Contributors*: Jonathan Mallard

File:Ancient Mechanical Artillery. Pic 01.jpg *Source*: http://en.wikipedia.org/w/index.php?title=File:Ancient_Mechanical_Artillery._Pic_01.jpg *License*: unknown *Contributors*: SBA73

File:French gunner 15th century - HHWXI264.svg *Source*: http://en.wikipedia.org/w/index.php?title=File:French_gunner_15th_century_-_HHWXI264.svg *License*: unknown *Contributors*: Mattes, Santosga, Struthious Bandersnatch, Vantey

File:View from sao jorge.JPG *Source*: http://en.wikipedia.org/w/index.php?title=File:View_from_sao_jorge.JPG *License*: unknown *Contributors*: Enrique Cornejo

File:HGM Pumhart von Steyr.jpg *Source*: http://en.wikipedia.org/w/index.php?title=File:HGM_Pumhart_von_Steyr.jpg *License*: unknown *Contributors*: Pappenheim

File:Zamek w Malborku - bombarda.jpg *Source*: http://en.wikipedia.org/w/index.php?title=File:Zamek_w_Malborku_-_bombarda.jpg *License*: unknown *Contributors*: Avron, Gruß Tom, Kresspahl, Ludmiła Pilecka, Palladinus

File:TheTsarCannonJuly2004.jpg *Source*: http://en.wikipedia.org/w/index.php?title=File:TheTsarCannonJuly2004.jpg *License*: unknown *Contributors*: Rcbutcher

File:Dardanelles Gun Turkish Bronze 15c.png *Source*: http://en.wikipedia.org/w/index.php?title=File:Dardanelles_Gun_Turkish_Bronze_15c.png *License*: unknown *Contributors*: Antur, Cplakidas, KTo288, Kurykh, Rcbutcher, Sémhur

File:Gabions.png *Source*: http://en.wikipedia.org/w/index.php?title=File:Gabions.png *License*: unknown *Contributors*: Basar, Buzz, Duk, Gruß Tom, JotaCartas, RobertLechner

File:Fortpoint02-01-2006.JPG *Source*: http://en.wikipedia.org/w/index.php?title=File:Fortpoint02-01-2006.JPG *License*: unknown *Contributors*: User:Miskatonic

File:Polish multiple gun from XVI-XVII century.png *Source*: http://en.wikipedia.org/w/index.php?title=File:Polish_multiple_gun_from_XVI-XVII_century.png *License*: unknown *Contributors*: Maciej Szczepańczyk

File:Bundesarchiv Bild 183-S36048, Frankreich, Arras, Artillerie.jpg *Source*: http://en.wikipedia.org/w/index.php?title=File:Bundesarchiv_Bild_183-S36048,_Frankreich,_Arras,_Artillerie.jpg *License*: unknown *Contributors*: Croquant, GGraf, Gorgo, Martin H., NSK Nikolaos S. Karastathis, Rcbutcher

File:Russian artillery fire in Berlin.jpg *Source*: http://en.wikipedia.org/w/index.php?title=File:Russian_artillery_fire_in_Berlin.jpg *License*: unknown *Contributors*: Alex Bakharev, PMG, Randbewohner, SuperTank17, Voevoda, 2 anonymous edits

File:WWI Field Gun 316.JPG *Source*: http://en.wikipedia.org/w/index.php?title=File:WWI_Field_Gun_316.JPG *License*: unknown *Contributors*: User:OlivierCotton

File:Nuclear artillery test Grable Event - Part of Operation Upshot-Knothole.jpg *Source*: http://en.wikipedia.org/w/index.php?title=File:Nuclear_artillery_test_Grable_Event_-_Part_of_Operation_Upshot-Knothole.jpg *License*: unknown *Contributors*: Federal government of the United States

File:152 mm howitzer D-20 belong to Military of Iran.jpg *Source*: http://en.wikipedia.org/w/index.php?title=File:152_mm_howitzer_D-20_belong_to_Military_of_Iran.jpg *License*: unknown *Contributors*: unknown

File:16-in Battleship Ammunition.JPG *Source*: http://en.wikipedia.org/w/index.php?title=File:16-in_Battleship_Ammunition.JPG *License*: unknown *Contributors*: U.S. Navy photo

File:Cyclone-French-320th-artillery-5 Sept-1917-Belgium.jpg *Source*: http://en.wikipedia.org/w/index.php?title=File:Cyclone-French-320th-artillery-5_Sept-1917-Belgium.jpg *License*: unknown *Contributors*: Bejnar, Hohum, Infrogmation, Rcbutcher, Schekinov Alexey Victorovich, Tony Rotondas

File:August 1984 - captured field guns in Jaji, Paktia.jpg *Source*: http://en.wikipedia.org/w/index.php?title=File:August_1984_-_captured_field_guns_in_Jaji,_Paktia.jpg *License*: unknown *Contributors*: Erwin Lux

File:3rdMarineDivisionVietnam1968h.JPEG *Source*: http://en.wikipedia.org/w/index.php?title=File:3rdMarineDivisionVietnam1968h.JPEG *License*: unknown *Contributors*: U.S Military photograph

File:PzH2000.Ger.jpg *Source*: http://en.wikipedia.org/w/index.php?title=File:PzH2000.Ger.jpg *License*: unknown *Contributors*: User:Sonaz

File:MuseeMarine-canonBronze-p1000434.jpg *Source*: http://en.wikipedia.org/w/index.php?title=File:MuseeMarine-canonBronze-p1000434.jpg *License*: unknown *Contributors*: User:Rama

File:Gasmasked artillerymen.jpg *Source*: http://en.wikipedia.org/w/index.php?title=File:Gasmasked_artillerymen.jpg *License*: unknown *Contributors*: Frank Hurley.

File:Tir.jpg *Source*: http://en.wikipedia.org/w/index.php?title=File:Tir.jpg *License*: unknown *Contributors*: Avron, Herrick, Kilom691, Rama, Rcbutcher, Topory, Traumrune

File:Artillery gun crew-illustration.jpeg *Source*: http://en.wikipedia.org/w/index.php?title=File:Artillery_gun_crew-illustration.jpeg *License*: unknown *Contributors*: Dodo, Foroa, Hohum, Jotsio

File:60 pounder Cape Helles June 1915.jpg *Source*: http://en.wikipedia.org/w/index.php?title=File:60_pounder_Cape_Helles_June_1915.jpg *License*: unknown *Contributors*: Gsl, Labattblueboy, Rcbutcher, Semnoz, Sevela.p, Shimgray, SoLando, Takabeg, Wikibob

File:DD-SD-00-00118.JPEG *Source*: http://en.wikipedia.org/w/index.php?title=File:DD-SD-00-00118.JPEG *License*: unknown *Contributors*: High Contrast, Rama

File:155fire.jpg *Source*: http://en.wikipedia.org/w/index.php?title=File:155fire.jpg *License*: unknown *Contributors*: DoD photo by Corporal Branden P. O'Brien, U.S. Marine Corps.

File:Gun5a.jpg *Source*: http://en.wikipedia.org/w/index.php?title=File:Gun5a.jpg *License*: unknown *Contributors*: Original uploader was Looper5920 at en.wikipedia

File:PzH2000Munition.jpg *Source*: http://en.wikipedia.org/w/index.php?title=File:PzH2000Munition.jpg *License*: unknown *Contributors*: User:KrisfromGermany

File:MRSI-trajectories.svg *Source*: http://en.wikipedia.org/w/index.php?title=File:MRSI-trajectories.svg *License*: unknown *Contributors*: User:Erzbischof

Image:Coat_of_arms_of_Sri_Lanka.svg *Source*: http://en.wikipedia.org/w/index.php?title=File:Coat_of_arms_of_Sri_Lanka.svg *License*: unknown *Contributors*: User:Tonyjeff

Image:Naval Ensign of Sri Lanka.svg *Source*: http://en.wikipedia.org/w/index.php?title=File:Naval_Ensign_of_Sri_Lanka.svg *License*: unknown *Contributors*: User:Denelson83

Image:Ensign of the Sri Lanka Air Force.svg *Source*: http://en.wikipedia.org/w/index.php?title=File:Ensign_of_the_Sri_Lanka_Air_Force.svg *License*: unknown *Contributors*: User:Thommy9

File:SLAC pic1.jpg *Source*: http://en.wikipedia.org/w/index.php?title=File:SLAC_pic1.jpg *License*: unknown *Contributors*: Geniac, Nitraven, We hope

File:Sri Lanka Air Force Kfir.jpg *Source*: http://en.wikipedia.org/w/index.php?title=File:Sri_Lanka_Air_Force_Kfir.jpg *License*: unknown *Contributors*: Gira2be (talk). Original uploader was Gira2be at en.wikipedia

MIX
Papier aus verantwortungsvollen Quellen
Paper from responsible sources
FSC® C105338

Printed by Books on Demand GmbH, Norderstedt / Germany